T0223461

Fourier Theory in Optics and Optical Information Processing

Multidisciplinary and Applied Optics

Series Editor:
Vasudevan Lakshminarayanan, University of Waterloo, Ontario, Canada

Quantum Mechanics of Charged Particle Beam Optics: Understanding Devices from Electron Microscopes to Particle Accelerators: Understanding Devices from Electron Microscopes to Particle Accelerators
Ramaswamy Jagannathan,
Sameen Ahmed Khan

Understanding Optics with Python
Vasudevan Lakshminarayanan, Hassen Ghalila, Ahmed Ammar, L. Srinivasa Varadharajan
Hassen Ghalila, University Tunis El Manar, Tunisia
Ahmed Ammar, University Tunis El Manar, Tunisia L.
Srinivasa Varadharajan, University of Waterloo, Ontario, Canada

Contemporary Holography
C. S. Narayanamurthy

Fourier Theory in Optics and Optical Information Processing
Toyohiko Yatagai, Utsunomiya University and University of Tsukuba

For more information about this series, please visit:
https://www.crcpress.com/Multidisciplinary-and-Applied-Optics/book-series/CRCMULAPPOPT

Fourier Theory in Optics and Optical Information Processing

Toyohiko Yatagai

CRC Press
Taylor & Francis Group
Boca Raton London New York

CRC Press is an imprint of the
Taylor & Francis Group, an **informa** business

First edition published 2022
by CRC Press
6000 Broken Sound Parkway NW, Suite 300, Boca Raton, FL 33487-2742

and by CRC Press
4 Park Square, Milton Park, Abingdon, Oxon, OX14 4RN

CRC Press is an imprint of Taylor & Francis Group, LLC

ISBN: 978-0-367-89457-3 (hbk)
ISBN: 978-0-367-64045-3 (pbk)
ISBN: 978-1-003-12191-6 (ebk)

DOI: 10.1201/9781003121916

Typeset in Nimbus
by KnowledgeWorks Global Ltd.

Publisher's note: This book has been prepared from camera-ready copy provided by the authors

Contents

Preface ix

Chapter 1   Light and Waves 1

    1.1   Waves and the Wave Equation 1
    1.2   Plane Wave 3
    1.3   Spherical Wave 6
    1.4   Complex Representation of Wave 7
    1.5   Principle of Superposition 8
    1.6   Scalar Wave and Vector Wave 9

Chapter 2   Interference and Diffraction 13

    2.1   Interference 13
    2.2   Fringe Visibility 15
    2.3   Young's Experiment 16
    2.4   Interferometer 18
    2.5   Diffraction 18
    2.6   Fresnel Diffraction 23
    2.7   Fraunhofer Diffraction 25
        2.7.1   Rectangular Aperture 27
        2.7.2   Circular Aperture 28
        2.7.3   Diffraction Grating 29

Chapter 3   Fourier Transform and Convolution 33

    3.1   Fourier Series 33
    3.2   Optimum Polynomial Approximation 39
    3.3   Normalized Orthogonal Polynomials 40
    3.4   Fourier Transform 41
    3.5   Some Representations of Fourier Transform 42
    3.6   Properties of the Fourier Transform 44
    3.7   Delta Function 47
    3.8   Convolution Integral and Correlation Function 49
    3.9   Some Functions and Their Fourier Transforms 52
    3.10 Sampling Theory 56

Chapter 4   Linear System 63

    4.1   System and Operator 63
    4.2   Linear System and Shift-Invariant System 64

    4.2.1 Linear System 64
    4.2.2 Shift-Invariant System 64
    4.2.3 Impulse Response 65
  4.3 Frequency Response Function 66
  4.4 Eigenfunction and Eigenvalue 67

Chapter 5 Discrete Fourier Transform and Fast Fourier Transform 71

  5.1 Discrete Fourier Transform 71
  5.2 Window Functions 73
  5.3 Principle of Fast Fourier Transform (FFT) 75
  5.4 Numerical Calculation Using FFT 78
  5.5 Interpolation in DFT 80
    5.5.1 Zero Padding 80
    5.5.2 Some Other Interpolation Techniques 82

Chapter 6 Fourier Optics 85

  6.1 Fresnel Diffraction 85
  6.2 Fourier Transform Operation of Lens 86
  6.3 Coherent Imaging 89
  6.4 Incoherent Imaging 93
  6.5 Frequency Response of Optical System 93
    6.5.1 Coherent Imaging 93
    6.5.2 Incoherent Imaging 94
  6.6 Resolving Power 95
  6.7 Angular Spectrum Method 96
  6.8 Diffraction Based on 3-D Fourier Spectrum 98

Chapter 7 Holography 105

  7.1 Conventional Optical Holography 105
  7.2 Computer Generated Holography 107
    7.2.1 Cell-Oriented CGH 107
    7.2.2 Point-Oriented CGH 109
    7.2.3 Kinoform 111
  7.3 Digital Holography 111

Chapter 8 Optical Computing 115

  8.1 Spatial Frequency Filtering 115
    8.1.1 Low-Pass and High-Pass Filters 116
    8.1.2 Differentiation and Laplacian Filters 117
    8.1.3 Phase-Contrast Filter 118
    8.1.4 Super Resolution and Apodization 119
  8.2 Matched Filter 120
  8.3 Optimum Filter for Additive Noise 123

8.4 Optimum Filter for Multiplicative Noise 125
8.5 Spectrum Analyzer 126
8.6 Optical Correlator 128
 8.6.1 Space-Integral Type 129
 8.6.2 Time-Integral Type 129
8.7 Joint Transform Correlator 130
8.8 Optical Addition and Optical Subtraction 131
8.9 Coordinate Transform 134
 8.9.1 Equal Magnification Imaging 135
 8.9.2 Logarithmic Coordinate Transform 136
8.10 Mellin Transform 136
8.11 Wavelet Transform 138
8.12 X-Ray Computer Tomography 140
 8.12.1 Two-Dimensional Fourier Transform Method 142
 8.12.2 Filtered Back Projection Method 142

Chapter 9   Analytic Signal and Hilbert Transform 147

9.1 Complex Representation and Negative Frequency 147
9.2 Analytic Signal 149
9.3 Hilbert Transform 150

Chapter 10   Coherence, Spectroscopy and Fringe Analysis 155

10.1 Coherence 155
 10.1.1 Temporal Coherence 157
 10.1.2 Spatial Coherence 159
10.2 Fourier Transform Spectroscopy 160
10.3 Phase Shift in Interferometry 163
10.4 Fourier Transform Fringe Analysis 167
10.5 Fringe Analysis by Hilbert Transform 168

Chapter 11   Spatio-Temporal Signal Processing 171

11.1 FemtoSecond Pulse Shaper 171
 11.1.1 Function of Grating 171
 11.1.2 Diffracted Beam 172
11.2 Spatial Frequency Filtering for Ultra-Short Pulse 174
11.3 Spatio-Temporal Joint Fourier Transform Correlator 176
11.4 Optical Coherence Tomography 179
11.5 Spectral Holography 182

Chapter 12   Wigner Distribution Function 185

12.1 WDF for Spatial Signal 185
 12.1.1 Definition and Its Properties 185
 12.1.2 WDF in Optical System 187

                    12.1.2.1   Lens Effect...187
                    12.1.2.2   Space Propagation 187
           12.2 WDF for Spatio-Temporal Signal......................................188
                12.2.1   Extension to Spatio-Temporal Signals..................188
                12.2.2   Lens Effect in Spatio-Temporal WDF 190
                12.2.3   Temporal Phase Modulator (Time Lens) 191
                12.2.4   Propagation and Dispersion 191
                12.2.5   Diffraction Grating...192
                12.2.6   Matrix Representation of WDF Transformation....193
                    12.2.6.1   Lens 193
                    12.2.6.2   Temporal Phase Modulation
                               (Time Lens) 193
                    12.2.6.3   Propagation and Dispersion...................194
                    12.2.6.4   Grating...194

Chapter 13   Fractional Fourier Transform 197

           13.1 Definition of Fractional Fourier Transform 197
           13.2 Some Representations of Fractional Fourier Transform.....197
           13.3 Applications to Optical Computing 201
                13.3.1   Wiener Filtering 201
                13.3.2   Correlator and Matched Filter................................202
                13.3.3   Joint Fractional Fourier Transform Correlator.......204

Appendix A   Numerical Calculation of Discrete Fresnel Diffraction...............211

Appendix B   Numerical Calculation of Fresnel Hologram 215

Solutions to Selected Problems 219

Index...235

Preface

Fourier analysis is one of the most important concepts in cases where you apply physical ideas to engineering issues. Optics, mechanics, electro-magnetism, quantum mechanics, signal processing, system and control theory, *etc.* are fundamental tools in physical engineering, in which Fourier analysis plays the most powerful roles and of which Fourier analysis is the most general concept.

The Fourier transform has been described fractionally in mathematics, physics, system theory, *etc.* in the past. Here I do try to straighten out the idea of Fourier transform again, since I think that a unified view of many phenomena in science and engineering can be obtained by understanding the concept of Fourier analysis. A deep understanding of one concept gives us great power for understanding new fields of science and engineering.

Fourier series building Fourier analysis was a mathematical means for solving thermal conduction problems, developed by J.B.J. Fourier. The Fourier transform is not only known as one of the very important mathematical methods but also is the basis of the response function and the power spectrum in the systems theory. This is an essential means for the systematic description of many phenomena in science and engineering. Fourier analysis is not an abstract concept in mathematics but it has a real physical meaning in many cases, for example, in diffraction and spectroscopy. In this case, the Fourier transform is directly observed as the intensity distribution.

In this book, optical phenomena are discussed based on Fourier analysis. I believe deep understanding of abstract meanings in Fourier analysis by optical phenomena gives us powerful tools for understanding all the concepts in science and engineering.

This book was originally published in 1992 in Japanese. It has been used as a textbook in many optical courses at graduate levels for more than two decades. Many students, optical researchers, and engineers who have read this book have been helpful in bettering it. I would like to express my sincere thanks to all of my former students, especially Prof. Boaz Jessie Jackin, Dr. Yusuke Sando and Prof. Ken-Ichiro Sugisaka, for improving the manuscript. In addition, I am especially grateful to Prof. Vasudevan Lakshminarayanan who suggested publishing this book and commented on the project.

Toyohiko Yatagai

1 Light and Waves

Light is an electromagnetic wave whose wavelengths are between that of radio waves and X-ray. The wavelength of light ranges from 100 μm to 10 nm and its frequency from 10^{12} Hz to 10^{17} Hz. In general, light is classified in terms of wavelength. The wave lengths of visible light expand between 380 nm for the violet and 800 nm for the red. The light with shorter wave length is called ultraviolet and longer one infrared. The visible spectrum of light is commonly formed as "light." In modern physics, light is understood as both a particle and a wave, but light is considered to be a wave in this book, since only the propagation phenomenon of light is mainly discussed. Particle property of light is manifested only in the case of energy interaction between electromagnetic waves and matter.

It should be noted that propagation laws and properties of light are valid over the entire range of the electromagnetic spectrum. We can discuss all phenomena, including the propagation of radio waves, imaging of visible light and X-ray diffraction, in a unified manner. This is the wave property of light.

1.1 WAVES AND THE WAVE EQUATION

Sound is the propagation of pressure variations or density changes of air. Light is the propagation of variation in the electric and magnetic fields.

Consider that the vibration u propagates in the z direction with the speed v, as shown in Fig. 1.1. Assuming the shape f of the wave variation u propagating in the z direction at time $t = 0$, we have

$$u(z, t = 0) = f(z). \tag{1.1}$$

The wave u moves a distance vt at time t but its shape is not changed, so we have

$$u(z,t) = f(z - vt). \tag{1.2}$$

This means that the wave variance does not change independently with variables of time t and position z, but only as a function of $z - vt$. The relationship among variation u, position z and time t exists but does not depend on the shape of variation f. Using

$$\tau = z - vt, \tag{1.3}$$

we have

$$\frac{\partial u}{\partial z} = \frac{\partial u}{\partial \tau} \cdot \frac{\partial \tau}{\partial z} = \frac{\partial u}{\partial \tau} \tag{1.4}$$

$$\frac{\partial u}{\partial t} = \frac{\partial u}{\partial \tau} \cdot \frac{\partial \tau}{\partial t} = -v \frac{\partial u}{\partial \tau}. \tag{1.5}$$

DOI: 10.1201/9781003121916-1

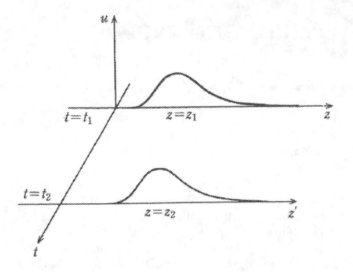

Figure 1.1   Propagation of wave.

Differentiating these again, we have

$$\frac{\partial^2 u}{\partial z^2} = \frac{\partial}{\partial \tau}\left(\frac{\partial u}{\partial \tau}\right)\frac{\partial \tau}{\partial z} = \frac{\partial^2 u}{\partial \tau^2} \tag{1.6}$$

$$\frac{\partial^2 u}{\partial t^2} = \frac{\partial}{\partial \tau}\left(\frac{\partial u}{\partial \tau}\right)\frac{\partial \tau}{\partial t} = v^2\frac{\partial^2 u}{\partial \tau^2}. \tag{1.7}$$

Therefore we have

$$\frac{\partial^2 u}{\partial z^2} = \frac{1}{v^2}\frac{\partial^2 u}{\partial t^2}. \tag{1.8}$$

This equation describes the wave propagating in the $+z$ direction with the velocity $\pm v$. This is called the wave equation.[1]

In general, by extending Eq. (1.8), the wave equation in three dimensions (x, y, z) can be written as

$$\frac{\partial^2 u}{\partial x^2} + \frac{\partial^2 u}{\partial y^2} + \frac{\partial^2 u}{\partial z^2} = \frac{1}{v^2}\frac{\partial^2 u}{\partial t^2}, \tag{1.9}$$

or by using the Laplacian operator

$$\nabla^2 = \frac{\partial^2}{\partial x^2} + \frac{\partial^2}{\partial y^2} + \frac{\partial^2}{\partial z^2}, \tag{1.10}$$

the 3-D wave equation is rewritten as

$$\nabla^2 u = \frac{1}{v^2}\frac{\partial^2 u}{\partial t^2}. \tag{1.11}$$

---

[1] The wave $f(z + vt)$ propagating to the $-z$ direction also satisfies Eq. (1.8).

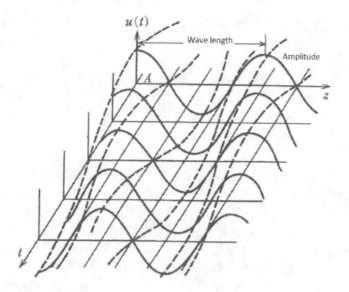

Figure 1.2 Periodicity of wave. Solid lines show sinusoidal waves in space and dotted lines show sinusoidal waves in time.

1.2 PLANE WAVE

The most simple solution of the wave equation is a sinusoidal wave. The sinusoidal wave propagation to the z direction with the velocity v is written as

$$u(z,t) = A\cos[k(z - vt) + \phi]. \qquad (1.12)$$

The equation evidently satisfies the wave equation (1.8). The maximum of variation A is called the amplitude, and $k(z - vt) + \phi$ is the phase. The sinusoidal wave is a periodic function both in space and time, as shown in Fig. 1.2. The period in space is called the wavelength, denoted by λ. If the wave propagates a distance of λ, the phase of the wave in Eq. (1.12) changes by 2π;

$$k\lambda = 2\pi, \qquad (1.13)$$

so we have

$$k = 2\pi/\lambda. \qquad (1.14)$$

Since k means a number of λ in the length of 2π and k is called the wave number or propagation constant, in the field of optical wave guides,[2] ϕ is initial phase and can be 0, if the spatial coordinate z and the time coordinate t are arbitrary values.

---

[2] In spectroscopy, $\sigma = 1/\lambda$ is called wave number. If both the wave numbers should be classified, k is called angular wave number and σ spectroscopic wave number.

The period T in time is given by

$$T = \lambda/v \qquad (1.15)$$

and the reciprocal of T is frequency

$$v = 1/T. \qquad (1.16)$$

Finally, we have the frequency

$$v = v/\lambda, \qquad (1.17)$$

which means frequency is a number of waves per unit distance of v (a propagation distance per unit time). The angular frequency is defined as

$$\omega = 2\pi v. \qquad (1.18)$$

The light velocity in vacuum is a physical constant c. In the case of light wave, v is light velocity in a medium. The ratio of v and c is refractive index n.

$$n = c/v. \qquad (1.19)$$

The frequency is written as

$$v = c/(n\lambda) = c/\lambda_0, \qquad (1.20)$$

where λ_0 denotes the light wavelength in vacuum. Therefore, λ is wavelength in a medium, which is written as

$$\lambda = \lambda_0/n. \qquad (1.21)$$

It should be noted that Eq. (1.12) is rewritten by

$$u(z,t) = A\cos(kz - \omega t), \qquad (1.22)$$

where the phase ϕ is omitted.

An equi-phase surface or a wavefront is characterized by the phase which remains constant at any time in 3-D space. Equations (1.12) and (1.22) describe a 1-D wave propagating to the z direction and also describe a wave in 3-D whose wavefront is planar and perpendicular to the z axis. The wave with planar wavefront is called a plane wave. It should be noted that a wave propagates to the direction perpendicular to its wavefront.

In order to describe a plane wave in 2-D or 3-D space, consider a sinusoidal plane wave propagating to the direction with an angle of θ to the x axis in the 2-D (x,y) coordinate system, as shown in Fig. 1.3. Since its wavefront is perpendicular to the propagation direction, this wave is written as

$$u(x,y,t) = A\cos(kX - \omega t), \qquad (1.23)$$

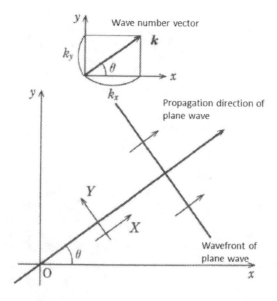

Figure 1.3   Plane wave propagating to the θ direction.

where the propagation direction is the direction of the X axis and the wavefront is on the Y axis. By coordinate transform from (X,Y) to (x,y),

$$u(x,y,t) = A\cos(k_x x + k_y y - \omega t),\qquad(1.24)$$

where

$$k_x = k\cos\theta\qquad(1.25)$$
$$k_y = k\sin\theta.\qquad(1.26)$$

Here, a wave number vector $\mathbf{k} = (k_x, k_y)$ is introduced. The direction of $\mathbf{k}$ is the propagation direction of the wave and $|\mathbf{k}| = k = 2\pi/\lambda$.

Extension of the 2-D wave propagation to 3-D give us a plane wave propagating in 3-D space written by

$$u(x,y,z,t) = A\cos(k_x x + k_y y + k_z z - \omega t),\qquad(1.27)$$

as shown in Fig. 1.4. In this case, the wave number vector is defined as $\mathbf{k} = (k_x, k_y, k_z)$. By introducing a position vector $\mathbf{r} = (x,y,z)$, an inner product is given by

$$\mathbf{k}\cdot\mathbf{r} = k_x x + k_y y + k_z z.\qquad(1.28)$$

Equation (1.27) is rewritten as

$$u(r,t) = A\cos(\mathbf{k}\cdot\mathbf{r} - \omega t).\qquad(1.29)$$

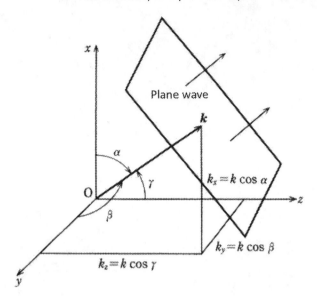

Figure 1.4 Plane wave propagating in 3-D space.

Let the directional cosines of $\boldsymbol{k}$ be $(\cos\alpha,\cos\beta,\cos\gamma)$. Equation (1.28) is rewritten as

$$\boldsymbol{k}\cdot\boldsymbol{r} = \frac{2\pi}{\lambda_0}n(x\cos\alpha + y\cos\beta + z\cos\gamma). \tag{1.30}$$

1.3 SPHERICAL WAVE

The wave with a spherical wavefront is called the spherical wave, as shown in Fig. 1.5. At first, consider the wave equation of Eq. (1.11) in spherical coordinates

$$\frac{1}{v^2}\frac{\partial^2 u}{\partial t^2} = \frac{1}{r}\frac{\partial^2(ru)}{\partial r^2}. \tag{1.31}$$

or equivalently,

$$\frac{1}{v^2}\frac{\partial^2(ru)}{\partial t^2} = \frac{\partial^2(ru)}{\partial r^2}. \tag{1.32}$$

The solution of this equation is

$$u(r,t) = \frac{1}{r}f(r\pm vt), \tag{1.33}$$

where $f(r-vt)/r$ represents a spherical wave expanding from the origin and $f(r+vt)/r$ a spherical wave converging toward the origin. The expanding sinusoidal spherical wave is given by

$$u(r,t) = \frac{A}{r}\cos\left(\frac{2\pi}{\lambda}r - \omega t\right). \tag{1.34}$$

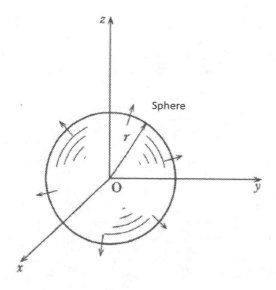

Figure 1.5 Spherical wave.

If the observation point locates far from the origin, the change of $1/r$ is not important and we have a simpler expression

$$u(r,t) = A\cos\left(\frac{2\pi}{\lambda}r - \omega t\right). \tag{1.35}$$

1.4 COMPLEX REPRESENTATION OF WAVE

In general, a plane wave is represented by Eq. (1.29), which is rewritten as

$$u(r,t) = \mathrm{Re}\{A\exp[\mathrm{i}(k\cdot r - \omega t)]\}, \tag{1.36}$$

because

$$\exp(\mathrm{i}\alpha) = \cos\alpha + \mathrm{i}\sin\alpha, \tag{1.37}$$

where $\mathrm{Re}\{...\}$ denotes the real part of complex number. Equation (1.36) is also represented by

$$u(r,t) = \frac{1}{2}\{A\exp[\mathrm{i}(k\cdot r - \omega t)] + A^*\exp[-\mathrm{i}(k\cdot r - \omega t)]\} \tag{1.38}$$

$$= \frac{1}{2}A\exp[\mathrm{i}(k\cdot r - \omega t)] + \mathrm{c.c.}, \tag{1.39}$$

where A^* denotes the complex conjugate of A, and c.c. means the complex conjugate of its former term. In some cases, the symbol of the real part $\mathrm{Re}\{...\}$ is omitted so that Eq. (1.36) is simply written as

$$u(r,t) = A\exp[\mathrm{i}(k\cdot r - \omega t)]. \tag{1.40}$$

This is called the complex amplitude. It should be noted that a wave that exists physically is represented by a real number and so the complex amplitude is only a mathematical expression. We have introduced the complex amplitude for mathematical convenience. For example, to calculate a sum of waves

$$A = \sum_m A_m \exp[i(k_m \cdot r - \omega t)] = \left[\sum_m A_m \exp(ik_m \cdot r)\right] \cdot \exp(-i\omega t), \quad (1.41)$$

we can separate a spatial part and a temporal part at first, and then calculate the spatial parts independently, and finally multiply the temporal part $\exp(-i\omega t)$. The real amplitude is given by the real part of the final result. In many optical cases, only the spatial terms are considered. If necessary, the time-dependent term is multiplied with the final results of spatial calculation. It should be noted that such methods in the complex notation of wave are valid only in the case of linear operations.

In a non-linear case, for example, to perform multiplication of waves, since

$$A_1 \exp(ik_1 \cdot r) \times A_2 \exp(ik_2 \cdot r) = A_1 A_2 \exp[i(k_1 + k_2) \cdot r], \quad (1.42)$$

its real part is $A_1 A_2 \cos[(k_1 + k_2) \cdot r]$, while the product of waves in real notation is $A_1 \cos(k_1 \cdot r) \times A_2 \cos(k_2 \cdot r)$. This means the product of waves is given by the product of the real parts of complex amplitudes but not by the real part of the product of complex amplitudes.

As is well known, the energy of the wave is given by the square of wave amplitude. Fortunately, the product of real parts of complex amplitudes is not necessary but the intensity of the complex amplitude is given by the square of the absolute value of the complex amplitude[3]

$$I = |u|^2. \quad (1.43)$$

1.5   PRINCIPLE OF SUPERPOSITION

Consider many waves coming to a point at the same time–for simplicity, two waves $f_1(z - vt)$ and $f_2(z - vt)$ satisfying the same wave equation.

$$\frac{\partial^2 f_1}{\partial z^2} = \frac{1}{v^2} \frac{\partial^2 f_1}{\partial t^2} \quad (1.44)$$

$$\frac{\partial^2 f_2}{\partial z^2} = \frac{1}{v^2} \frac{\partial^2 f_2}{\partial t^2}. \quad (1.45)$$

The composite wave f is given by the sum of amplitudes of the two waves f_1 and f_2.

$$f = f_1 + f_2, \quad (1.46)$$

---

[3]The intensity of the light is defined as the energy crossing a unit area perpendicularly in unit time. In the electromagnetic theory of light, the instantaneous energy is given by the Poynting vector S. Because the frequency of the light is extremely high, only the time-averaged value of the magnitude of the Poynting vector $\langle S \rangle$ can be measured by photo-detectors. This gives $\langle \varepsilon v |E|^2 \rangle / 2$, where ε, v, E and $\langle \cdots \rangle$ denote the dielectric constant of media, the light velocity in the media, the electric field of light and the time-averaging, respectively. In optics, the square of the electric field of light gives the light intensity.

By using Eq. (1.8),

$$\frac{\partial^2 f}{\partial z^2} = \frac{\partial^2 (f_1 + f_2)}{\partial z^2} = \frac{\partial^2 f_1}{\partial z^2} + \frac{\partial^2 f_2}{\partial z^2}$$

$$= \frac{1}{v^2}\frac{\partial^2 f_1}{\partial t^2} + \frac{1}{v^2}\frac{\partial^2 f_2}{\partial t^2} = \frac{1}{v^2}\frac{\partial^2 (f_1 + f_2)}{\partial t^2} = \frac{1}{v^2}\frac{\partial^2 f}{\partial t^2}. \tag{1.47}$$

This means the composite wave satisfies the wave equation. That is, the sum of amplitudes of waves gives the amplitude of the composite wave. This is *the principle of superposition* in wave physics. The principle of superposition is one of the most fundamental principles in wave physics, which is due to the linearity of the wave equation. An example is shown in Fig. 1.6 where two waves are propagating to the opposite direction to each other.

The principle of superposition claims that the sum of waves can exist as a wave. On the contrary, one wave can be decomposed into a sum of arbitrary numbers of waves. The example of Eq. (1.46) implies that the wave f is decomposed into two waves f_1 and f_2.

There are many ways of decomposing waves. The most useful one is the decomposition of plane waves with different wave numbers.

$$f(z - vt) = A_0 + A_1 \cos[k(z - vt) + \phi_1] + A_2 \cos[2k(z - vt) + \phi_2] + \cdots$$

$$= \sum_m A_m \cos[mk(z - vt) + \phi_m], \tag{1.48}$$

where A_m denotes the amplitude of a decomposed plane wave. The coefficient A_m is the degree of the contribution of the m-th decomposed plane wave for the original wave f. It should be noted that all the summed waves are solutions of the wave equation that the original wave satisfies.

The mathematical condition for possible decomposition will be discussed later in detail but it is good to note here that this decomposition is called a Fourier series.

1.6 SCALAR WAVE AND VECTOR WAVE

Until now, we did not consider the direction of the electric field variation u. The light is an electromagnetic wave. Assuming the light is propagating to the direction of the z axis, the variation direction of the electric and magnetic fields are the directions of the x and y axes. This type of wave is called a transverse wave. On the other hand, an acoustic wave is a longitudinal wave, where the direction of variation is in the propagation direction z.

In general, the electric field and the magnetic field are vectors with three components: $E(E_x, E_y, E_z)$ and $H(H_x, H_y, H_z)$, respectively. Therefore, the light wave propagation is vertical by nature.

In a homogeneous media, like vacuum, water or glass, the optical properties do not depend on the position and the propagation direction, and hence the components $E_x, E_y, E_z, H_x, H_y, H_z$ satisfy the wave equation independently:

$$\nabla^2 E_x = \frac{1}{v^2}\frac{\partial^2 E_x}{\partial t^2}, \tag{1.49}$$

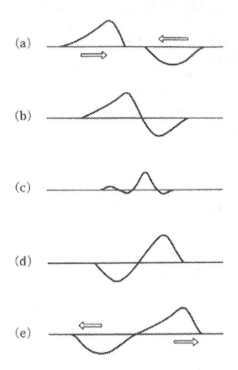

Figure 1.6 Collision of waves.

and so on. Those equations are integrated into the wave equation Eq. (1.11). This wave is called the scalar wave.

Generally, the light wave is considered as a scalar wave, but in an inhomogeneous media or near an aperture or boundary of homogeneous media, the components of electric and magnetic fields are not independent and interact with each other. In such a case, the scalar approximation is not valid and the light wave should be considered as a vector wave.

Next, consider a complex sinusoidal wave as a solution of scalar wave equation,

$$u(r,t) = U(r)\exp(-i\omega t), \tag{1.50}$$

where

$$U(r) = A(r)\exp[i\phi(r)]. \tag{1.51}$$

Since this equation satisfies the wave equation (1.11), substituting Eq. (1.50) into Eq. (1.11) gives the Helmholtz equation

$$(\nabla^2 + k^2)U = 0, \tag{1.52}$$

where k denotes the wave number (1.14). The Helmholtz equation Eq. (1.52) describes the monochromatic wave propagation in a homogeneous medium.

PROBLEMS

1. Show that a wave $u = f(z + vt)$ propagating to the $-z$ direction satisfies the wave equation (Eq. (1.11)).
2. Consider the plane wave described by

$$u(t) = 20\cos[2\pi(32t - 8z)]. \qquad (1.53)$$

   Evaluate its velocity, direction of propagation, period, wavelength and wave number, where the unit of length and time are meters and seconds.
3. Evaluate the frequency and wave number of a light wave with a wavelength of $0.6328\ \mu$m.
4. Consider that two waves

$$u_1 = A\cos[2\pi/\lambda_1(z - vt)]$$

and

$$u_2 = A\cos[2\pi/\lambda_2(z - vt)]$$

with the wavelengths λ_1 and λ_2, respectively, are slightly different. Draw the compound wave of the above two waves in time $t = 0$ and $t = 1$ s, where $A = 2.0$ m, $v = 3.0$ m/s, $\lambda_1 = 8.0$ m, $\lambda_2 = 6.0$ m.

BIBLIOGRAPHY

Hecht, E. 2002. *Optics*, 4th ed. Pearson Education.
Born, M. and Wolf, E. 1999. *Principles of Optics*, 7th ed. Cambridge University Press.

2 Interference and Diffraction

In this chapter, the phenomena of interference and diffraction are discussed, and some distinctive properties of the light wave are presented. The superposition of light waves constructs the phenomenon of interference and diffraction. At first, the interference of two plane waves and its fringe visibility is discussed. The fringe visibility is related with the coherence. Then Young's experiment and some interferometers are introduced. As an extension of Young's experiment, the diffraction is introduced. In particular, the Fraunhofer diffraction can be represented as the Fourier transform. Some numerical examples of diffraction patterns of simples objects are presented. These examples will give an intuitive understanding of the Fourier transform.

2.1 INTERFERENCE

Consider two sinusoidal waves passing from point A to point C and from point B to point C. For simplicity, the frequencies of the two waves are the same. Vectors of AC and BC are denoted by $\mathbf{r}_{AC}$ and $\mathbf{r}_{BC}$, respectively, and their wave number vectors by $\mathbf{k}_A$ and $\mathbf{k}_B$, respectively (see Fig. 2.1). The plane wave arriving at point C from point A is given by

$$u_A = A_A \exp[i(\mathbf{k}_A \cdot \mathbf{r}_{AC} + \phi_A - \omega t)], \tag{2.1}$$

and the plane wave arriving at point C from point B is

$$u_B = A_B \exp[i(\mathbf{k}_B \cdot \mathbf{r}_{BC} + \phi_B - \omega t)]. \tag{2.2}$$

Suppose the position vectors of points A and C are $\mathbf{r}_A(x_A, y_A, z_A)$ and $\mathbf{r}_C(x_C, y_C, z_C)$, we have

$$\mathbf{r}_{AC} = \mathbf{r}_C - \mathbf{r}_A. \tag{2.3}$$

By using Eq. (1.30), we have

$$\mathbf{k}_A \cdot \mathbf{r}_{AC} = \mathbf{k} \cdot (\mathbf{r}_C - \mathbf{r}_A)$$
$$= \frac{2\pi}{\lambda_0} n[(x_C - x_A) \cos \alpha_A + (y_C - y_A) \cos \beta_A + (z_C - z_A) \cos \gamma_A], \tag{2.4}$$

where the directional cosines of the vector $\mathbf{k}_A$ are $(\cos \alpha_A, \cos \beta_A, \cos \gamma_A)$. If

$$(x_C - x_A) \cos \alpha_A + (y_C - y_A) \cos \beta_A + (z_C - z_A) \cos \gamma_A = l_{AC}, \tag{2.5}$$

then l_{AC} gives the distance between points A and C, and $n l_{AC}$ is called the optical distance, where n is the refractive index of the medium. Equations (2.1) and (2.2) are rewritten as

$$u_A = A_A \exp \left[i \left(\frac{2\pi}{\lambda_0} n l_{AC} + \phi_A - \omega t \right) \right] \tag{2.6}$$

DOI: 10.1201/9781003121916-2

13

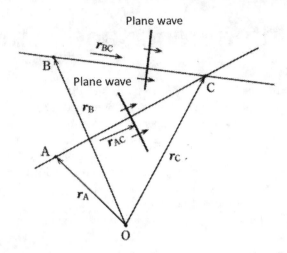

Figure 2.1 Interference of plane waves.

and

$$u_B = A_B \exp\left[i\left(\frac{2\pi}{\lambda_0}nl_{BC} + \phi_B - \omega t\right)\right], \tag{2.7}$$

respectively. Since the amplitude u_C at point C is given by the superposition of two waves u_A and u_B, we have

$$u_C = A_A \exp\left[i\left(\frac{2\pi}{\lambda_0}nl_{AC} + \phi_A - \omega t\right)\right] + A_B \exp\left[i\left(\frac{2\pi}{\lambda_0}nl_{BC} + \phi_B - \omega t\right)\right]$$

$$= \left\{A_A \exp\left[i\left(\frac{2\pi}{\lambda_0}nl_{AC} + \phi_A\right)\right] + A_B \exp\left[i\left(\frac{2\pi}{\lambda_0}nl_{BC} + \phi_B\right)\right]\right\} \exp(-i\omega t). \tag{2.8}$$

Therefore, the intensity at point C is given by

$$I_C = |u_C|^2$$

$$= I_A + I_B + 2\sqrt{I_A I_B} \cos\left[\frac{2\pi}{\lambda_0}n(l_{BC} - l_{AC}) + (\phi_B - \phi_A)\right], \tag{2.9}$$

where $I_A = |u_A|^2$ and $I_B = |u_B|^2$.

The intensity I_C is not the summation of the intensity I_A and I_B, but the third term is appended. This term is the term of interference, and it changes sinusoidally according to the optical path difference of $n(l_{BC} - l_{AC})$, and the fringe pattern as shown in Fig. 2.2 is generated. If the phase difference is defined as

$$\Phi = \frac{2\pi}{\lambda_0}n(l_{BC} - l_{AC}) + \phi_B - \phi_A, \tag{2.10}$$

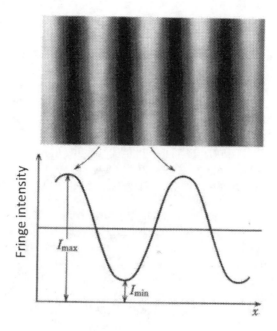

Figure 2.2 Interference fringes (superposition of two plane waves).

the maximum I_{max} and the minimum I_{min} of the fringe intensity are given by

$$I_{\text{max}} = (A_A + A_B)^2 \qquad : \Phi = 2\pi m \qquad\qquad (2.11)$$

$$I_{\text{min}} = (A_A - A_B)^2 \qquad : \Phi = \pi(2m+1), \qquad\quad (2.12)$$

where m is an integer. The time-dependent term in Eq. (2.9) disappears and the interference fringe intensity is independent of time. This means the time term can be neglected in the beginning equations of Eqs. (2.1) and (2.2) in the case of the interference of waves with the same frequency. In this book, we omit the time-dependent terms unless otherwise required.

2.2 FRINGE VISIBILITY

As the measure of interference fringe clarity, the contrast or the visibility is defined by

$$V = \frac{I_{max} - I_{min}}{I_{max} + I_{min}}, \qquad\qquad (2.13)$$

where I_{max} and I_{min} are the maximum and the minimum of the fringe intensity, respectively. We have

$$V = \frac{2\sqrt{I_A I_B}}{I_A + I_B} \qquad\qquad (2.14)$$

by using Eq. (2.9). The maximum contrast of $V = 1$ is obtained when $I_A = I_B$. When either I_A or I_B is 0, the minimum $V = 0$ so that the fringe is invisible.

Whenever $A_A = A_B$, is the contrast always $V = 1$? In reality, a special condition is necessary for $V = 1$. The interference fringe exists stably in time, only when the difference $\phi_B - \phi_A$ of the initial phase of ϕ_A and ϕ_B at the point A and B is stable in time. The difference $\phi_B - \phi_A$ depends on the properties of the light sources, the distances from the light source to the point A and B, and so on. This means the contrast of the interference fringe depends not only on the path difference of $l_{BC} - l_{AC}$ but also the light source properties and the layout of the optical system.

The phase of wave from the light source in many cases is stable in less than 10^{-8} seconds. We can see the wave shape is sinusoidal only within this short time, where the amplitude and the phase are fixed within an appropriate time. Many wavelets with fixed duration generated from the light source form practical waves. From this concept, to generate a stable interference fringe at point C, the waves A and B departing from the same source and at the same time should superimpose at point C. The device used to perform such a superposition is called as an interferometer.

As described in Section 10.1, the stability of the phase difference $\phi_B - \phi_A$ depends on the properties of the light source. The measure is called the degree of coherence, γ_{AB}. We have $0 \leq \gamma_{AB} \leq 1$. In the case of $\gamma_{AB} = 1$, two waves from the point A and B are called coherent each other, and incoherent in the case of $\gamma_{AB} = 0$. When considering the coherence, the fringe contract is rewritten as

$$V = \frac{2\sqrt{I_A I_B}}{I_A + I_B} \gamma_{AB}. \tag{2.15}$$

Therefore Eq. (2.9) describes the fringe intensity in the case of the coherent light source $\gamma_{AB} = 1$.

The fringe intensity equation (2.9) is rewritten as

$$I_C = I_A + I_B + 2\sqrt{I_A I_B} \gamma_{AB} \cos\left[\frac{2\pi}{\lambda_0} n(l_{BC} - l_{AC}) + (\phi_B - \phi_A)\right]. \tag{2.16}$$

In the case of incoherent waves,

$$I_C = I_A + I_B, \tag{2.17}$$

the sum of the intensity I_A and I_B gives the total intensity of I_C. In the case of coherent waves, the intensity of the fringe is obtained by the square of the sum of the amplitudes of waves, as shown in Eq. (2.9).

2.3　YOUNG'S EXPERIMENT

Consider a simple optical arrangement for observing the optical interference, as shown in Fig. 2.3. This is the most famous optical interferometer developed by Thomas Young in 1801 to show that light is a wave. Two narrow slits are located behind the single slit of the light source. On the observation plane located at a sufficient distance from the double slits, the waves from slits A and B are superposed.

Suppose the plane of the double slit and the screen plane are parallel to each other, and its distance is R. The origin O of the coordinates is set at the center of the double slit. The light source, the slit S and the origin O are located on the axis of z. The double slits are located at $x = d/2$ and $x = -d/2$. The slit width is very small as the order of the light wavelength λ, and therefore the waves after the slits propagate as expanding spherical waves in the x-z plane. On the screen, two waves from the slits A and B are superimposed. Because the distances from the slit S to A and B are equal, the initial phases ϕ_A and ϕ_B of the spherical waves are always equal and coherent with each other.

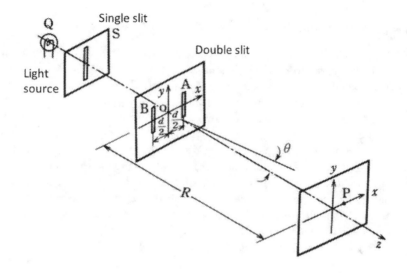

Figure 2.3 Optical setup of Young's experiment.

The wave propagating from slit A to the point P on the observation plane is given by

$$u_A = A \exp(ikr_{AP}),\tag{2.18}$$

in the case when the distance R is sufficiently large, $R \gg d/2$ and $R \gg x$, where

$$r_{AP} = \sqrt{R^2 + \left(x - \frac{d}{2}\right)^2} \simeq R + \frac{(x - d/2)^2}{2R}.\tag{2.19}$$

Similarly, the spherical wave propagating from the slit B to the observation point P on the screen is given by

$$u_B = A \exp(ikr_{BP}),\tag{2.20}$$

where

$$r_{BP} = \sqrt{R^2 + \left(x + \frac{d}{2}\right)^2} \simeq R + \frac{(x + d/2)^2}{2R}.\tag{2.21}$$

Finally, the intensity of the interference fringe at P is given by

$$\begin{aligned}
I_P &= |u_A + u_B|^2 \\
&= 2A^2[1 + \cos k(r_{BP} - r_{AP})] \\
&= 2A^2\left[1 + \cos\left(\frac{kd}{R}x\right)\right].
\end{aligned}$$

(2.22)

A sinusoidal fringe with an equal period is observed on the screen. This is called the Young fringe, whose contrast $V = 1$.

2.4   INTERFEROMETER

The interferometer is an equipment that makes interference of waves by dividing and superimposing waves from a single light source. A variety of interferometers have been developed in order to measure distances, refractive indices, and so on. They are measurement tools employing a precise scale of the order of the wavelength of light. Commonly used interferometers are shown in Fig. 2.4.

The Michelson interferometer Fig. 2.4(a) is of historical importance, which was used for measuring the length of the metric standard and also showing that no ether exists. The Twyman-Green interferometer Fig. 2.4(b) and the Fizeau interferometer Fig. 2.4(c) are used for surface shape measurement in optics and precision machining. The Mach-Zehnder interferometer Fig. 2.4(d) is used for visualization of flows and index distribution of gas and plasma. The refractive index of gas and liquid is measured by the Jamin interferometer Fig. 2.4(e). The Fabry-Perot interferometer Fig. 2.4(f) is used as a precise spectrometer.

2.5   DIFFRACTION

Consider the Young interferometer shown in Fig. 2.3 again. For simplicity of analysis, assume that the single slit S locates at a sufficiently long distance from the double slits A and B so that the waves from slits A and B can be considered to be plane waves. The screen plane is also at a sufficiently long distance. The angle between the point x and the origin on the screen plane is given by

$$\theta = \frac{x}{R}.$$

(2.23)

The intensity of the Young fringe on the screen plane is written as a function of the angle θ,

$$\begin{aligned}
I &= |u_A + u_B|^2 = \left|A\exp\left(-i\frac{kd\theta}{2}\right) + A\exp\left(i\frac{kd\theta}{2}\right)\right|^2 \\
&= 2A^2[1 + \cos(kd\theta)],
\end{aligned}$$

(2.24)

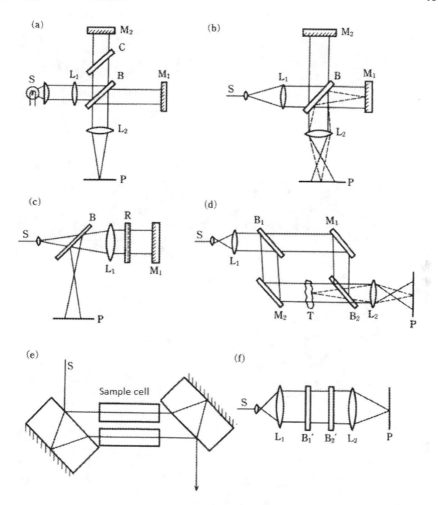

Figure 2.4 Typical interferometers. (a) Michelson interferometer, (b) Twyman-Green interferometer, (c) Fizeau interferometer, (d) Mach-Zehnder interferometer, (e) Jamin interferometer, and (f) Fabry-Perot interferometer. S: light source, L: lens, M: mirror, B: beam splitter, C: compensator, R: reference surface, B': high-reflective surface, T: transparent object, P: observation plane.

where u_A and u_B are given approximately by

$$u_A = A \exp\left(-i\frac{kd\theta}{2}\right) \cdot \exp(ikR) \tag{2.25}$$

$$u_B = A \exp\left(i\frac{kd\theta}{2}\right) \cdot \exp(ikR). \tag{2.26}$$

The fringe profile is shown in Fig. 2.5(a).

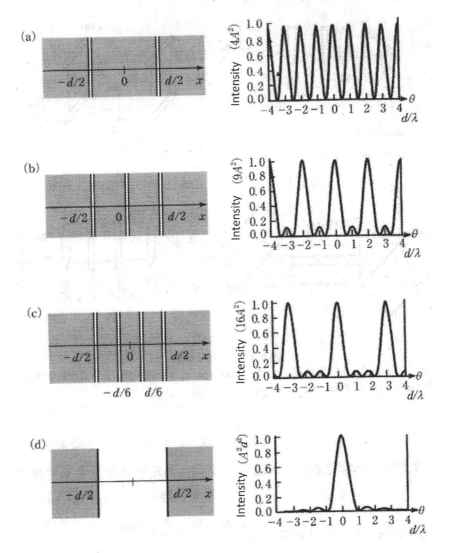

Figure 2.5 Diffraction of multi-slits (a)–(c) and a single slit (d).

Then consider the three slit case as shown in Fig. 2.5(b). In analogy with the double slit case, we have the fringe intensity for the three slit case as

$$I = \left| A\exp\left(-i\frac{kd\theta}{2}\right) + A + A\exp\left(i\frac{kd\theta}{2}\right) \right|^2$$
$$= 3A^2 + 4A^2\cos\left(\frac{kd\theta}{2}\right) + 2A^2\cos(kd\theta). \tag{2.27}$$

In the case of four slits as shown in Fig. 2.5(c), we have

$$I = \left| A\exp\left(-i\frac{kd\theta}{2}\right) + A\exp\left(-i\frac{kd\theta}{6}\right) + A\exp\left(i\frac{kd\theta}{6}\right) + A\exp\left(i\frac{kd\theta}{2}\right) \right|^2$$

$$= 4A^2 + 6A^2\cos\left(\frac{kd\theta}{3}\right) + 4A^2\cos\left(\frac{2kd\theta}{3}\right) + 2A^2\cos(kd\theta). \tag{2.28}$$

Using a similar procedure, we have an aperture with the width of d by increasing the number of slits but without changing the width of d. In this case, the amplitude distribution on the screen plane is given by

$$u = \int_{-d/2}^{d/2} A\exp(-ik\theta\tau)d\tau = Ad\frac{\sin(kd\theta/2)}{kd\theta/2}. \tag{2.29}$$

Then the intensity distribution is given by

$$I = A^2 d^2 \left[\frac{\sin(kd\theta/2)}{kd\theta/2}\right]^2. \tag{2.30}$$

The wave expands on a screen located far from the aperture, which is illuminated by a coherent light. The diffraction is an optical phenomenon of the wave propagation after an obstacle such as an aperture, where the wave propagates with some divergence. To define the extent of the diffraction pattern, we consider the distance from the center of the diffraction pattern to the first zero intensity,

$$\frac{kd\theta}{2} = \pi. \tag{2.31}$$

Then the angular extension is given by

$$\theta = \frac{\lambda}{d}. \tag{2.32}$$

The extent of the diffraction is proportional to the wavelength and inversely proportional to the size of the aperture. As shown in Fig. 2.6, due to the diffraction, the wave propagates behind the obstacle. The shorter the width of the aperture, more obvious is the diffraction effect, and therefore, the part of straight propagation is reduced. If the width of the aperture is the same as the wavelength, then the diffracted wave can be to be considered as a spherical wave. Therefore, in Sec. 2.3, the waves from slits were regarded as spherical waves. The integral of Eq. (2.29) means that the superposition of many spherical waves[1] originating at the aperture plane gives the amplitude of the diffracted wave.

In the history of optics, the Huygens principle shown in Fig. 2.7 has described diffraction. According to the Huygens principle, every point on the incident wavefront Σ at a time is a source of secondary waves, called wavelets. The envelope Σ' of wavelets results in the new wavefront in the next time. This principle does not explain the diffraction completely. Fresnel described the interference of wavelets gives the amplitude of the diffracted wave. This is also called the Huygens principle.

---

[1]$A\exp(-ik\theta\tau)$ in Eq. (2.29) is the plane wave but this is due to a long propagation distance from the slit plane to the observation point.

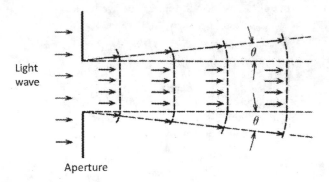

Light
wave

Aperture

Figure 2.6   Diffraction from aperture.

According to the Huygens principle, the diffraction of Fig. 2.5(d) is formulated as

$$u_P = \iint_S \frac{A \exp(ikr)}{i\lambda r} dxdy, \tag{2.33}$$

by using coordinates shown in Fig. 2.8, where r denotes the distance from a point in the aperture to the observation point P, $dxdy$ a small area element on the aperture plane and the integral is over the area of the aperture.

The wavelet is described by $A/r \exp(ikr)$. Assume that the distance R from the aperture plane to the observation plane is sufficiently larger than the aperture size. In this case, the change of r is very small if the position (x,y) is changed inside the aperture, and therefore $1/r$ is considered to be constant and it is equal to $1/R$.

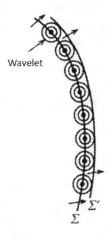

Wavelet

Figure 2.7   Huygens principle.

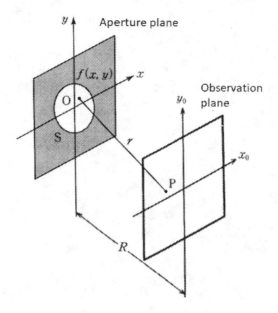

Figure 2.8 Coordinates for diffraction formulation.

Next, we define the aperture function,

$$f(x,y) = \begin{cases} 1: & \text{inside aperture } S \\ 0: & \text{outside aperture } S. \end{cases} \qquad (2.34)$$

Equation (2.33) is rewritten by

$$u_P(x,y) = \frac{A}{i\lambda R} \iint_{-\infty}^{\infty} f(x,y)\exp(ikr)dxdy. \qquad (2.35)$$

This equation gives us the amplitude of the diffracted wave from the aperture S illuminated by a plane wave normal to the aperture plane.

2.6 FRESNEL DIFFRACTION

For further discussion, a simplified version of Eq. (2.35) is derived. The distance from a point (x,y) in the aperture to the observation point $P(x_o,y_o)$ is given by

$$r = \sqrt{R^2 + (x-x_o)^2 + (y-y_o)^2}. \qquad (2.36)$$

Since $R \gg (x - x_o)^2$ and $(y - y_o)^2$, Eq. (2.36) can be approximated as

$$r = R\sqrt{1 + \frac{(x - x_o)^2 + (y - y_o)^2}{R^2}}$$
$$\approx R + \frac{1}{2R}[(x - x_o)^2 + (y - y_o)^2] - \frac{1}{8R^3}[(x - x_o)^2 + (y - y_o)^2]^2. \quad (2.37)$$

Here we suppose

$$\left| \frac{k}{8R^3}[(x - x_o)^2 + (y - y_o)^2]^2 \right| \ll 1. \quad (2.38)$$

In other words, if

$$R^3 \gg \frac{\pi}{4\lambda}[(x - x_o)^2 + (y - y_o)^2]^2, \quad (2.39)$$

Eq. (2.37) is rewritten by

$$r = R + \frac{1}{2R}[(x - x_o)^2 + (y - y_o)^2]. \quad (2.40)$$

Therefore, the diffraction equation (2.35) is given by

$$u_P(x_o, y_o) = \frac{A}{i\lambda R} \exp(ikR) \iint_{-\infty}^{\infty} f(x, y) \exp\left\{ \frac{ik}{2R}[(x - x_o)^2 + (y - y_o)^2] \right\} dxdy. \quad (2.41)$$

The diffraction under the condition of Eq. (2.39) is called the Fresnel diffraction and Eq. (2.41) is called the Fresnel diffraction equation.

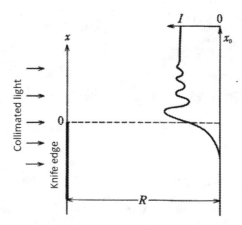

Figure 2.9 Geometry of Fresnel diffraction from a knife-edge.

Consider the simplest case, light passing through a knife-edge, as shown in Fig. 2.9. Here the obstacle object, knife-edge is located at $x < 0$. The Fresnel diffraction equation for the above condition is given by

$$u_P(x_o) = \frac{A}{i\lambda R} \exp(ikR) \int_0^\infty \exp\left[i\frac{\pi}{\lambda R}(x-x_o)^2\right] dx. \tag{2.42}$$

This equation is modified as

$$u_P(x_o) = \frac{A}{i} \sqrt{\frac{1}{2\lambda R}} \exp(ikR) \int_{\xi'}^\infty \exp\left(i\frac{\pi}{2}\xi^2\right) d\xi \tag{2.43}$$

with

$$\xi = \sqrt{\frac{2}{\lambda R}}(x-x_o), \tag{2.44}$$

where the lower limit of the integral is

$$\xi' = -\sqrt{\frac{2}{\lambda R}} x_0. \tag{2.45}$$

By using Fresnel's integrals,

$$C(\xi) = \int_0^\xi \cos\left(\frac{\pi}{2}\xi^2\right) d\xi \tag{2.46}$$

$$S(\xi) = \int_0^\xi \sin\left(\frac{\pi}{2}\xi^2\right) d\xi, \tag{2.47}$$

Eq. (2.43) is finally rewritten by

$$u_P(x_o) = \frac{A}{i} \sqrt{\frac{1}{2R\lambda}} \exp(ikR)\{C(\infty) - C(\xi') + i[S(\infty) - S(\xi')]\}. \tag{2.48}$$

The Fresnel diffraction of Eq. (2.48) results in Fresnel integrals, which is not an elementary function. The value of the Fresnel integrals is obtained numerically.

The intensity of the diffraction pattern is given by

$$I_P(x_o) \propto [C(\infty) - C(\xi')]^2 + [S(\infty) - S(\xi')]^2. \tag{2.49}$$

As shown in Fig. 2.10, the diffracted wave propagates into the area of the geometrical shadow and visible fringes appear in an unobstructed area.

Figure 2.11 shows the intensities of the Fresnel diffraction patterns of a slit at different observation points.

2.7 FRAUNHOFER DIFFRACTION

Further approximation of Eq. (2.40) gives us

$$r = R - \frac{1}{R}(xx_o + yy_o) + \frac{1}{2R}(x_o^2 + y_o^2) \tag{2.50}$$

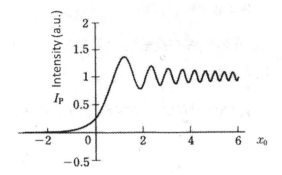

Figure 2.10 Fresnel diffraction pattern of a knife-edge.

by neglecting squared terms of x^2 and y^2, under the condition

$$\left| \frac{k}{2R}(x^2 + y^2) \right| \ll 1, \tag{2.51}$$

which is more restricted than the Fresnel diffraction case. The condition is given by

$$R \gg \frac{\pi}{\lambda}(x^2 + y^2). \tag{2.52}$$

For example, $R > 1.2$ km in the case when the aperture width is 1 cm and the wavelength 0.5 μm.

By substituting Eq. (2.50) to Eq. (2.35), we have

$$u_P(x_o, y_o) = \frac{A}{i\lambda R} \exp(ikR) \cdot \exp\left[i\frac{k}{2R}(x_o^2 + y_o^2)\right]$$
$$\times \iint f(x, y) \exp\left[-i\frac{k}{R}(xx_o + yy_o)\right] dxdy. \tag{2.53}$$

We define

$$v_x = \frac{x_o}{\lambda R}, \quad v_y = \frac{y_o}{\lambda R} \tag{2.54}$$

then we have

$$u_P(v_x, v_y) = A' \iint f(x, y) \exp[-i2\pi(xv_x + yv_y)] dxdy, \tag{2.55}$$

where

$$A' = \frac{A}{i\lambda R} \exp(ikR) \cdot \exp\left[i\frac{k}{2R}(x_o^2 + y_o^2)\right]. \tag{2.56}$$

Equation (2.53) or Eq. (2.55) under the condition of Eq. (2.52) is called Fraunhofer diffraction.

The integral of Eq. (2.53) is also known as the Fourier integral or Fourier transform. The objective of this book is to discuss the physical and engineering meaning of Fourier transform in terms of optics. But in this chapter, we consider Eq. (2.53) as the diffraction integral and discuss this integral with some examples.

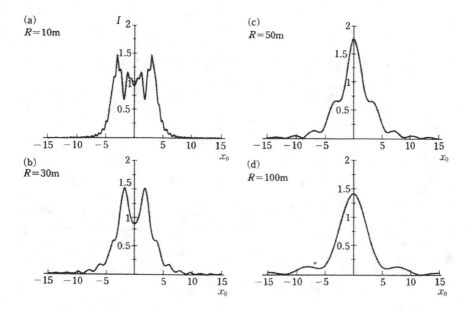

Figure 2.11 Fresnel diffraction patterns of a slit. The slit width is 10 mm and $\lambda = 0.63$ μm.

2.7.1 RECTANGULAR APERTURE

Consider the Fraunhofer diffraction from a rectangular aperture with the size of $D_x \times D_y$. From Eq. (2.53), the amplitude of the diffraction is given by

$$u(x_o, y_o) = A' \int_{-D_y/2}^{D_y/2} \int_{-D_x/2}^{D_x/2} \exp\left[-i\frac{k}{R}(xx_o + yy_o)\right] dxdy$$

$$= A'D_xD_y \frac{\sin\left(\frac{kD_x}{2R}x_o\right)}{\frac{kD_x}{2R}x_o} \cdot \frac{\sin\left(\frac{kD_y}{2R}y_o\right)}{\frac{kD_y}{2R}y_o}. \qquad (2.57)$$

By using the sinc function

$$\text{sinc}(x) = \frac{\sin \pi x}{\pi x}, \qquad (2.58)$$

Eq. (2.57) is rewritten as

$$u(x_o, y_o) = A'D_xD_y\text{sinc}\left(\frac{D_xx_0}{\lambda R}\right)\text{sinc}\left(\frac{D_yy_0}{\lambda R}\right). \qquad (2.59)$$

Therefore, its intensity distribution is

$$I(x_o, y_o) = A'^2 D_x^2 D_y^2 \text{sinc}^2\left(\frac{D_xx_0}{\lambda R}\right)\text{sinc}^2\left(\frac{D_yy_0}{\lambda R}\right). \qquad (2.60)$$

The amplitude distribution of the Fraunhofer diffraction of a rectangular aperture is shown in Fig. 2.12, and its intensity distribution in Fig. 2.13.

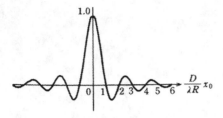

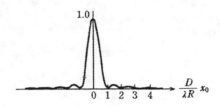

Figure 2.12 Amplitude of Fraunhofer diffraction of a rectangular aperture.

Figure 2.13 Intensity of Fraunhofer diffraction of a rectangular aperture.

2.7.2 CIRCULAR APERTURE

Consider the Fraunhofer diffraction of a circular aperture with its diameter D. In polar coordinates,

$$x = \rho \cos \theta, \quad y = \rho \sin \theta$$
$$x_o = w \cos \phi, \quad y_o = w \sin \phi. \tag{2.61}$$

Eq. (2.53) is rewritten as

$$u(w, \phi) = A' \int_0^{D/2} \int_0^{2\pi} \exp\left[-i\frac{k}{R}\rho w \cos(\theta - \phi)\right] \rho \,d\rho \,d\theta. \tag{2.62}$$

Using the n-th order of the first kind of Bessel function,

$$J_n(x) = \frac{i^{-n}}{2\pi} \int_0^{2\pi} \exp(ix\cos\alpha) \cdot \exp(in\alpha)\,d\alpha, \tag{2.63}$$

we have

$$u(w) = 2\pi A' \int_0^{D/2} J_0\left(\frac{k}{R}\rho w\right) \rho \,d\rho. \tag{2.64}$$

Using the Bessel function relationship

$$\frac{d}{dx}[x^{n+1}J_{n+1}(x)] = x^{n+1}J_n(x), \tag{2.65}$$

Eq. (2.64) is rewritten as

$$u(w) = \pi A'\left(\frac{D}{2}\right)^2 \cdot \frac{2J_1\left(\frac{kD}{2R}w\right)}{\frac{kD}{2R}w}. \tag{2.66}$$

Finally, the intensity of the Fraunhofer diffraction of a circular aperture is

$$I(w) = I_0\left[\frac{2J_1\left(\frac{kD}{2R}w\right)}{\frac{kD}{2R}w}\right]^2, \tag{2.67}$$

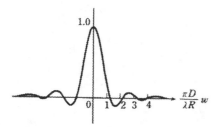

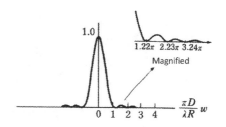

Figure 2.14 Amplitude of Fraunhofer diffraction of a circular aperture.

Figure 2.15 Intensity of Fraunhofer diffraction of a circular aperture.

where

$$I_0 = \frac{\pi^2 A'^2 D^4}{16}. \tag{2.68}$$

The amplitude distribution of the Fraunhofer diffraction of a circular aperture and its intensity distribution are shown in Figs. (2.14) and (2.15), respectively.

The intensity minimum is zero at 1.22π, 2.233π, 3.238π, and so on for the value of $kDw/2R$. The second maximum of the intensity is at $kDw/2R = 1.635$, where $I/I_0 = 0.0175$. The energy of the diffraction pattern is concentrated in the central disc area, which is called the Airy disc. The size of the Airy disc is estimated by the radius Δw of the first dark ring, we have

$$\Delta w = 1.22\frac{\lambda R}{D}. \tag{2.69}$$

2.7.3 DIFFRACTION GRATING

Consider the Fraunhofer diffraction of a grating consisting of straight N slits with a width of w parallel to each other, as shown in Fig. 2.16. Let the spacing of slits be a.

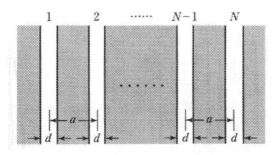

Figure 2.16 Diffraction grating.

Referring to Eq. (2.55), we have the diffracted wave from the n-th slit

$$u_n(v_x) = A' \int_{an-d/2}^{an+d/2} \exp(-i2\pi x v_x) dx$$

$$= A' \exp(-i2\pi a n v_x) \int_{-d/2}^{d/2} \exp(-i2\pi x v_x) dx$$

$$= A' \exp(-i2\pi a n v_x) u_0(v_x), \tag{2.70}$$

where

$$u_0(v_x) = \int_{-d/2}^{d/2} \exp(-i2\pi x v_x) dx. = d\mathrm{sinc}(dv_x) \tag{2.71}$$

The total diffracted wave is given by

$$u(v_x) = \sum_{n=0}^{N-1} u_n(v_x)$$

$$= A' u_o(v_x) \frac{1 - \exp(-i2\pi a N v_x)}{1 - \exp(-i2\pi a v_x)}. \tag{2.72}$$

The intensity of the Fraunhofer diffraction of a grating is given by

$$I(v_x) = I_0 \mathrm{sinc}^2(dv_x) \frac{1 - \cos(2\pi a N v_x)}{1 - \cos(2\pi a v_x)}$$

$$= I_0 \mathrm{sinc}^2(dv_x) \cdot U(v_x), \tag{2.73}$$

where

$$I_0 = A'^2 d^2 \tag{2.74}$$

and

$$U(v_x) = \left[\frac{\sin(\pi a N v_x)}{\sin(\pi a v_x)} \right]^2. \tag{2.75}$$

The Fraunhofer diffraction of a grating is a multiple of two parts, the diffraction from a single slit, Eq. (2.71) and a factor due to N slits, Eq. (2.75). The factor of Eq. (2.75) is the maximum of N^2 for the value

$$\pi a v_x = \pi n \quad (n : \text{integer}) \tag{2.76}$$

and zero for the value

$$\pi a N v x = \pi m \quad (m : \text{integer, not for multiple of } N) \tag{2.77}$$

as shown in Fig. 2.17(a), consisting of sharp peaks.

The intensity distribution of the Fraunhofer diffraction of a grating is shown in Fig 2.17(b), which is a multiple of the factor of Eq. (2.75) and a diffraction of a single slit of Eq. (2.71). The bright spot located on the optical axis is called the zero-order diffraction spot, and spots located outside of the zero-order are $\pm$ first, $\pm$ second, $\cdots$ diffraction orders. The width of the diffraction spot decreases as the number N of the slits increases. The width of the envelope of the diffraction pattern (this is due to the diffraction of a single slit) becomes wider as the width of the slits become shorter.

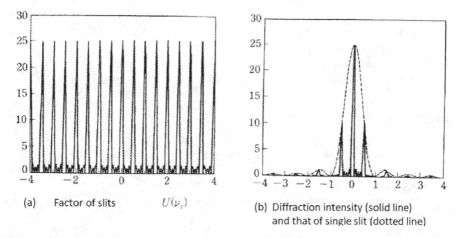

(a)   Factor of slits $U(\nu_x)$

(b) Diffraction intensity (solid line)
     and that of single slit (dotted line)

Figure 2.17 Fraunhofer diffraction of grating. (a) Factor of $U(\nu_x)$, and (b) diffraction intensity from grating (solid line) and from a single slit (dotted line).

PROBLEMS

1. Explain what kinds of phenomena happen when two waves with different frequencies are superimposed.
2. Two slits of the width of D located at a distance of l, illuminated by a monochromatic light of the wavelength λ. Calculate the Fraunhofer diffraction pattern, where $D < l$.
3. Calculate the Fraunhofer diffraction pattern of an annular aperture whose outer and inner diameters are D_1 and D_2, illuminated with a monochromatic light of the wavelength λ.

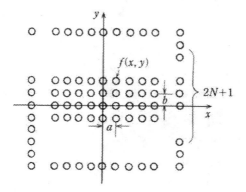

Figure 2.18 Aperture array.

4. Calculate the Fraunhofer diffraction pattern of the aperture array as shown in Fig. 2.18, where all the apertures are same described by $f(x,y)$, the number of apertures is $(2M+1) \times (2N+1)$, and the spacing to the x- and y-directions are a and b, respectively.

What is the diffraction pattern when apertures are distributed randomly?

BIBLIOGRAPHY

Hecht, E. 2002. *Optics*, 4th ed. Pearson Education.

Born, M. and Wolf, E. 1999. *Principles of Optics*, 7th ed. Cambridge University Press.

Guenther, R. 1990. *Modern Optics*. John Wiley & Sons., New York.

Sharma, K. K. 2006. *Optics*. Academic Press, Amsterdam.

Lakshminarayanan, K. K. and Varadharajan, L. S. 2015. *Special Functions for Optical Science and Engineering*. SPIE Press.

3 Fourier Transform and Convolution

The Fourier transform plays a very important role in a multitude of scientific and engineering fields. Fourier analysis is a universal tool for obtaining a number of possible solutions, which are suitable from a broad view of science and technology.

In this chapter, we describe the concept of the Fourier series at first and then its extension, the Fourier transform, is introduced. To apply Fourier analysis to optics, we do not emphasize mathematical rigor but rather the intuitive and conceptual understanding of Fourier analysis. Next, we discuss an important concept in system analysis, that is, the convolution integral, the correlation function and their Fourier transforms. Finally, the sampling theorem is introduced to transform analog signals into digital signals. Mathematical prerequisites in this book are completed by this description.

3.1 FOURIER SERIES

The sum of sinusoidal functions $\cos(2\pi n x/T + \phi_n), (n = 0, 1, 2, ...)$ with weights A_n

$$f(x) = \sum_{n=0}^{\infty} A_n \cos\left(2\pi \frac{n}{T} x + \phi_n\right) \tag{3.1}$$

is called as Fourier series. To make symmetric expression, Eq. (3.1) is rewritten as

$$f(x) = \frac{a_0}{2} + \sum_{n=1}^{\infty} \left[a_n \cos\left(2\pi \frac{n}{T} x\right) + b_n \sin\left(2\pi \frac{n}{T} x\right)\right], \tag{3.2}$$

where

$$a_0 = 2A_0 \cos \phi_0 \tag{3.3}$$

$$a_n = A_n \cos \phi_n \tag{3.4}$$

$$b_n = -A_n \sin \phi_n. \tag{3.5}$$

The right-hand side of Eqs. (3.1) and (3.2) are all periodic functions, and therefore, the function $f(x)$ is also a periodic function.

In Sec. 1.5, we described that Fourier series can be interpreted as expansion of a periodic function $f(x)$ into sinusoidal functions. The weight factors in Eq. (3.2) are

DOI: 10.1201/9781003121916-3

a_n and b_n. The weights can be obtained as follows; at first, both terms in Eq. (3.2) are multiplied by $\cos(2\pi mx/T)$ and then integrated in a period $[-T/2, T/2]$.

$$\int_{-T/2}^{T/2} f(x) \cos\left(2\pi\frac{m}{T}x\right) dx = \int_{-T/2}^{T/2} \frac{a_0}{2} \cos\left(2\pi\frac{m}{T}x\right) dx$$

$$+ \sum_{n=1}^{\infty} a_n \cos\left(2\pi\frac{n}{T}x\right) \cos\left(2\pi\frac{m}{T}x\right) dx$$

$$+ \sum_{n=1}^{\infty} b_n \sin\left(2\pi\frac{n}{T}x\right) \cos\left(2\pi\frac{m}{T}x\right) dx. \qquad (3.6)$$

Using the orthogonality of sinusoidal functions

$$\int_{-T/2}^{T/2} \cos\left(2\pi\frac{n}{T}x\right) \cos\left(2\pi\frac{m}{T}x\right) dx = \begin{cases} T/2 & (n=m) \\ 0 & (n \neq m) \end{cases} \qquad (3.7)$$

$$\int_{-T/2}^{T/2} \sin\left(2\pi\frac{n}{T}x\right) \cos\left(2\pi\frac{m}{T}x\right) dx = 0 \qquad (3.8)$$

$$\int_{-T/2}^{T/2} \sin\left(2\pi\frac{n}{T}x\right) \sin\left(2\pi\frac{m}{T}x\right) dx = \begin{cases} T/2 & (n=m) \\ 0 & (n \neq m) \end{cases} \qquad (3.9)$$

we have

$$a_n = \frac{2}{T} \int_{-T/2}^{T/2} f(x) \cos\left(2\pi\frac{m}{T}x\right) dx \quad (n = 0, 1, 2, ...). \qquad (3.10)$$

Similarly, the both terms in Eq. (3.2) are multiplied by $\sin(2\pi mx/T)$ and then integrated in a period $[-T/2, T/2]$, we have

$$b_n = \frac{2}{T} \int_{-T/2}^{T/2} f(x) \sin\left(2\pi\frac{n}{T}x\right) dx \quad (n = 1, 2, 3, ...). \qquad (3.11)$$

Here we used the sinusoidal function orthogonality of Eqs. (3.8) and (3.9).

Figure 3.1 shows the procedure for obtaining coefficients a_n for a periodic rectangular function with a width of d. Fourier series coefficients a_n give amounts of how many components of $\cos(2\pi nx/T)$ are included in a function $f(x)$ within the period T. Since $\cos(2\pi x/T)$ includes one period within T, $\cos(2\pi 2x/T)$ two periods, and in general, $\cos(2\pi nx/T)$ n periods, n is called the frequency, which is the same as the definition in Eq. (1.16). The coefficients a_n and b_n are the measure of frequency components, called the spectrum. This means that the spectrum of a periodic function $f(x)$ is a set of discrete frequencies a_n and b_n (n is an integer).

Just in the case when the wave is described by a complex exponential function, a complex version of Fourier series is convenient in many cases.

$$f(x) = \sum_{n=-\infty}^{\infty} c_n \exp\left(i2\pi\frac{n}{T}x\right), \qquad (3.12)$$

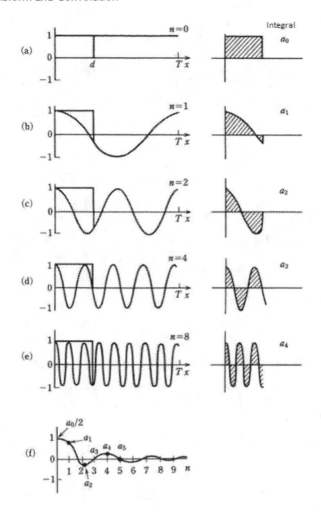

Figure 3.1 Calculation of Fourier coefficient a_n.

where

$$c_0 = \frac{a_0}{2} \tag{3.13}$$

$$c_n = \frac{a_n - ib_n}{2} \tag{3.14}$$

$$c_{-n} = \frac{a_n + ib_n}{2} = c_n^* \tag{3.15}$$

and c_n^* denotes a complex conjugate of c_n. From Eqs.(3.10) and (3.11), we have

$$c_n = c_{-n}^* = \frac{1}{T} \int_{-T/2}^{T/2} f(x) \exp\left(-i2\pi \frac{n}{T} x\right) dx. \tag{3.16}$$

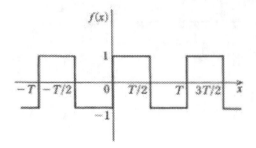

Figure 3.2 Periodic function $f(x) = -1(-T/2 \le x < 0)$, and $= 1(0 \le x < T/2)$.

Figure 3.2 shows a periodic function with a period of T, which is defined as

$$f(x) = \begin{cases} -1 & (-T/2 \le x < 0) \\ 1 & (0 \le x < T/2). \end{cases} \tag{3.17}$$

Using Eqs. (3.10) and (3.11), we have

$$a_n = \frac{2}{T}\left[\int_{-T/2}^{0}(-1)\cos\left(2\pi\frac{nx}{T}\right)dx + \int_{0}^{T/2}(1)\cos\left(2\pi\frac{nx}{T}\right)dx\right] = 0 \tag{3.18}$$

$$b_n = \frac{2}{T}\left[\int_{-T/2}^{0}(-1)\sin\left(2\pi\frac{nx}{T}\right)dx + \int_{0}^{T/2}(1)\sin\left(2\pi\frac{nx}{T}\right)dx\right]$$
$$= \frac{4}{\pi}\cdot\frac{1}{n} \quad (n : \text{odd integer}). \tag{3.19}$$

Finally, we have

$$f(x) = \frac{4}{\pi}\left(\sin\frac{2\pi x}{T} + \frac{1}{3}\sin\frac{6\pi x}{T} + \frac{1}{5}\sin\frac{10\pi x}{T} + \cdots\right) \tag{3.20}$$

whose series up to $n = 5$ is shown in Fig. 3.3.

The function shown in Fig. 3.4 is written as

$$f(x) = \begin{cases} x+1 & (-T/2 \le x < 0) \\ -x+1 & (0 \le x < T/2) \end{cases} \tag{3.21}$$

and its Fourier series coefficients are

$$a_0 = 2 - \frac{T}{2} \tag{3.22}$$

$$a_n = \frac{T}{\pi^2 n^2}[1 - (-1)^n] \tag{3.23}$$

$$b_n = 0. \tag{3.24}$$

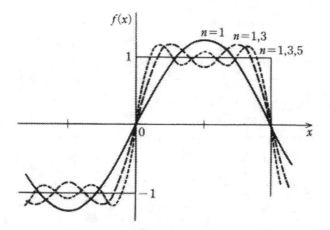

Figure 3.3 Shape of the periodic function shown in Fig. 3.2 and its Fourier series ($n \leq 5$).

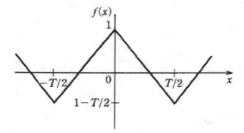

Figure 3.4 Periodic triangle function.

Finally, we have

$$f(x) = \frac{1}{2}\left(2 - \frac{T}{2}\right) + \frac{2T}{\pi^2}\left(\cos\frac{2\pi}{T}x + \frac{1}{3^2}\cos\frac{6\pi}{T}x + \frac{1}{5^2}\cos\frac{10\pi}{T}x + \cdots\right). \quad (3.25)$$

Typical periodic functions and their Fourier series are shown in Table 3.1.

The Fourier series of Eqs. (3.20) and (3.25), either a_n or b_n is zero. This is due to whether the function $f(x)$ is an odd function or an even function. A function $f(x)$ such that $f(x) = f(-x)$ for all x is called an even function. A function such that $f(x) = -f(-x)$ is an odd function. The Fourier series coefficient a_n relates to the cosine (even function) components of the function $f(x)$ and b_n their sine (odd function) components. It is easily understood that $a_n = 0$ if the function $f(x)$ is an odd function and $b_n = 0$ if the function $f(x)$ is an even function.

Table 3.1
Periodic Functions and Their Fourier Series

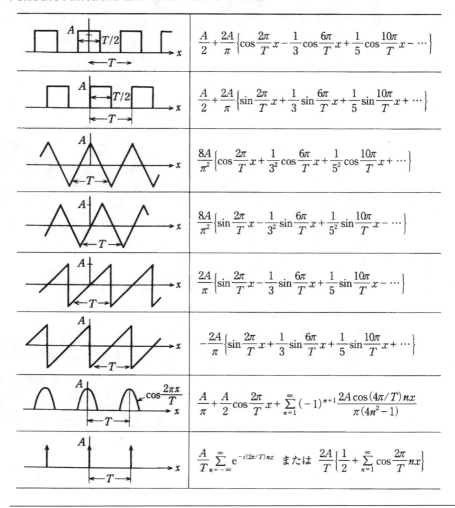

	$\dfrac{A}{2} + \dfrac{2A}{\pi}\left\{\cos\dfrac{2\pi}{T}x - \dfrac{1}{3}\cos\dfrac{6\pi}{T}x + \dfrac{1}{5}\cos\dfrac{10\pi}{T}x - \cdots\right\}$
	$\dfrac{A}{2} + \dfrac{2A}{\pi}\left\{\sin\dfrac{2\pi}{T}x + \dfrac{1}{3}\sin\dfrac{6\pi}{T}x + \dfrac{1}{5}\sin\dfrac{10\pi}{T}x + \cdots\right\}$
	$\dfrac{8A}{\pi^2}\left\{\cos\dfrac{2\pi}{T}x + \dfrac{1}{3^2}\cos\dfrac{6\pi}{T}x + \dfrac{1}{5^2}\cos\dfrac{10\pi}{T}x + \cdots\right\}$
	$\dfrac{8A}{\pi^2}\left\{\sin\dfrac{2\pi}{T}x - \dfrac{1}{3^2}\sin\dfrac{6\pi}{T}x + \dfrac{1}{5^2}\sin\dfrac{10\pi}{T}x - \cdots\right\}$
	$\dfrac{2A}{\pi}\left\{\sin\dfrac{2\pi}{T}x - \dfrac{1}{3}\sin\dfrac{6\pi}{T}x + \dfrac{1}{5}\sin\dfrac{10\pi}{T}x - \cdots\right\}$
	$-\dfrac{2A}{\pi}\left\{\sin\dfrac{2\pi}{T}x + \dfrac{1}{3}\sin\dfrac{6\pi}{T}x + \dfrac{1}{5}\sin\dfrac{10\pi}{T}x + \cdots\right\}$
	$\dfrac{A}{\pi} + \dfrac{A}{2}\cos\dfrac{2\pi}{T}x + \displaystyle\sum_{n=1}^{\infty}(-1)^{n+1}\dfrac{2A\cos(4\pi/T)nx}{\pi(4n^2-1)}$
	$\dfrac{A}{T}\displaystyle\sum_{n=-\infty}^{\infty}e^{-i(2\pi/T)nx}\quad\text{または}\quad\dfrac{2A}{T}\left\{\dfrac{1}{2}+\displaystyle\sum_{n=1}^{\infty}\cos\dfrac{2\pi}{T}nx\right\}$

In general, any function $f(x)$ can be split into an even function $f_e(x)$ and an odd function $f_o(x)$, which are

$$f_e(x) = \frac{f(x) + f(-x)}{2} \tag{3.26}$$

$$f_o(x) = \frac{f(x) - f(-x)}{2}. \tag{3.27}$$

Of course, we have

$$f(x) = f_e(x) + f_o(x). \tag{3.28}$$

Fourier series coefficients are given by

$$a_n = \frac{4}{T} \int_0^{T/2} f_e(x) \cos\left(\frac{2\pi n}{T} x\right) dx \tag{3.29}$$

$$b_n = \frac{4}{T} \int_0^{T/2} f_o(x) \sin\left(\frac{2\pi n}{T} x\right) dx. \tag{3.30}$$

It should be noted that Fourier series for an arbitrary function $f(x)$ does not always exist. For a Fourier series to exist, the function $f(x)$ should be finite and integrable within the interval $[-T/2, T/2]$, but not continuous. As shown in Fig.3.5, if a function $f(x)$ and its derivative $f'(x)$ are continuous except for a countable number of discontinuities, the function $f(x)$ is called a partially smooth function. Provided that a partially smooth function has a discontinuity at $x = x_1$, its Fourier series is proved to converges to the average $[f(x_1 - 0) + f(x_1 + 0)]/2$, where $f(x_1 - 0)$ and $f(x_1 + 0)$ are the left-hand and right-hand limits at the discontinuity $x = x_1$, respectively.

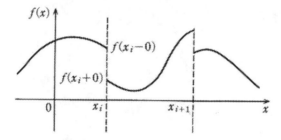

Figure 3.5 Partially smooth function.

3.2 OPTIMUM POLYNOMIAL APPROXIMATION

Consider a periodic function $f(x)$ with a period T. Then this function is approximated by a polynomial of sinusoidal functions,

$$P_N(x) = \frac{\alpha_0}{2} + \sum_{n=1}^{N} \left[\alpha_n \cos\left(\frac{2\pi nx}{T}\right) + \beta_n \sin\left(\frac{2\pi nx}{T}\right)\right]. \tag{3.31}$$

The merit function of a root-mean-square error is introduced to evaluate the approximation,

$$Q_N = \frac{1}{T} \int_{-T/2}^{T/2} \left[f(x) - P_N(x)\right]^2 dx. \tag{3.32}$$

Next, we find α_n and β_n to minimize the merit function. This is called as the optimum polynomial approximation. At first, Eq. (3.32) is expanded,

$$Q_N = \frac{1}{T} \int_{-T/2}^{T/2} f(x)^2 dx - \frac{2}{T} \int_{-T/2}^{T/2} f(x) P_N(x) dx + \frac{1}{T} \int_{-T/2}^{T/2} P_N(x)^2 dx. \tag{3.33}$$

Substituting Eq. (3.31) into Eq. (3.33), we have

$$
\begin{aligned}
Q_n =\ & \frac{1}{T} \int_{-T/2}^{T/2} f(x)^2 dx \\
& -\frac{2}{T} \left\{ \frac{\alpha_0}{2} \int_{-T/2}^{T/2} f(x) dx + \sum_{n=1}^{N} \left[\alpha_n \int_{-T/2}^{T/2} f(x) \cos\left(\frac{2\pi nx}{T}\right) dx \right. \right. \\
& \left. \left. +\beta_n \int_{-T/2}^{T/2} f(x) \sin\left(\frac{2\pi nx}{T}\right) dx \right] \right\} + \frac{1}{T} \int_{-T/2}^{T/2} \frac{\alpha_0^2}{4} dx \\
& +\frac{\alpha_0}{T} \sum_{n=1}^{N} \int_{-T/2}^{T/2} \left[\alpha_n \cos\left(\frac{2\pi nx}{T}\right) + \beta_n \sin\left(\frac{2\pi nx}{T}\right) \right] dx \\
& +\frac{1}{T} \sum_{n=1}^{N} \int_{-T/2}^{T/2} \left[\left(\alpha_n \cos\left(\frac{2\pi nx}{T}\right) + \beta_n \sin\left(\frac{2\pi nx}{T}\right) \right) \right]^2 dx.
\end{aligned} \tag{3.34}
$$

To obtain α_0, α_n and β_n for minimizing Q_n, we set $\partial Q_N/\partial \alpha_0 = 0$, $\partial Q_N/\partial \alpha_n = 0$ and $\partial Q_N/\partial \beta_n = 0$. Finally, we have

$$
\alpha_0 = \frac{2}{T} \int_{-T/2}^{T/2} f(x) dx \tag{3.35}
$$

$$
\alpha_n = \frac{2}{T} \int_{-T/2}^{T/2} f(x) \cos\left(\frac{2\pi nx}{T}\right) dx \tag{3.36}
$$

$$
\beta_n = \frac{2}{T} \int_{-T/2}^{T/2} f(x) \sin\left(\frac{2\pi nx}{T}\right) dx. \tag{3.37}
$$

Coefficients in Eqs. (3.35) to (3.37) are equal to that of Eqs. (3.10) and (3.11). When a periodic function $f(x)$ is approximated by the n-th order sinusoidal polynomial $P_N(x)$, the optimum coefficients of the polynomial are the coefficients of the Fourier series. It should be noted that the coefficients α_n and β_n of the optimum polynomial are uniquely determined, not depending on the order of N. When the approximation order is increased from N-th to $N + 1$-th, the old coefficients α_n and β_n are able to be employed as they are. Only new coefficients α_{n+1} and β_{n+1} should be calculated.

3.3 NORMALIZED ORTHOGONAL POLYNOMIALS

Coefficients c_n in complex Fourier series of Eq. (3.12) are calculated by using Eq. (3.16), but we can calculate directly c_n with the property of Fourier polynomials

$$
\frac{1}{T} \int_{-T/2}^{T/2} \exp\left(-i\frac{2\pi nx}{T} \right) \cdot \exp\left(i\frac{2\pi mx}{T} \right) dx = \delta_{n,m}, \tag{3.38}
$$

where $\delta_{n,m}$ is the Kronecker delta defined by

$$
\delta_{n,m} = \begin{cases} 1 & (n = m) \\ 0 & (n \neq m). \end{cases} \tag{3.39}
$$

This means that the Fourier coefficient c_n in the complex series is obtained by multiplying $\exp(-i2\pi nx/T)$ to the both sides of Eq. (3.12) and integrating it within the period $[-T/2, T/2]$. The property of Eq. (3.38) enables us to calculate the Fourier coefficient c_n in the complex series. The polynomials expanded by $\exp(i2\pi nx/T)$, (n: integer) have this property. This property is called orthogonality. Polynomials $\exp(i2\pi nx/T)$ are orthogonal within the period $[-T/2, T/2]$.

In general, consider the functions $\{f_n(x)\}$ and their inner product is defined by

$$(f_n, f_m) = \int_{-T/2}^{T/2} f_n(x) \cdot f_m^*(x) \, dx \tag{3.40}$$

and * denotes the complex conjugate. When

$$(f_n, f_m) = \delta_{n,m}, \tag{3.41}$$

where functions f_n and f_m are said to be orthogonal to each other and $\{f_n(x)\}$ are orthogonal functions. The norm of the function $f(x)$ is defined by

$$\| f \| = (f, f)^{1/2} \tag{3.42}$$

and the normalization of the function $f(x)$ is performed by multiplying an arbitrary constant to the function $f(x)$ so that its norm is unity. The normalized and orthogonal polynomials are called the normalized orthogonal polynomials. For example, the following polynomials are normalized orthogonal polynomials.

$$\sqrt{\frac{1}{T}} \exp\left(i\frac{2\pi nx}{T}\right), \quad n = 0, \pm 1, \pm 2, \dots \tag{3.43}$$

$$\sqrt{\frac{2}{T}} \cos\left(i\frac{2\pi nx}{T}\right), \quad n = 0, 1, 2, \dots \tag{3.44}$$

$$\sqrt{\frac{2}{T}} \sin\left(i\frac{2\pi nx}{T}\right), \quad n = 1, 2, 3, \dots \tag{3.45}$$

3.4   FOURIER TRANSFORM

As mentioned in Sec. 3.1, a periodic function can be expanded by the sum of sinusoidal functions, that is, Fourier series. What about non-periodic functions? As a matter of fact, non-periodic functions can be expanded in a similar manner. This is the Fourier transform. This extension is easy. As shown in Fig. 3.6, it should be noted that a periodic function becomes a non-periodic function when the period T increases to infinity $T \to \infty$. Consider a periodic function $f(x)$ with the period T expanded as a complex Fourier series,

$$f(x) = \sum_{n=-\infty}^{\infty} c_n \exp\left(i\frac{2\pi nx}{T}\right) \tag{3.46}$$

$$c_n = \frac{1}{T} \int_{-T/2}^{T/2} f(x) \exp\left(-i\frac{2\pi nx}{T}\right) dx. \tag{3.47}$$

Then c_n is regarded as a function of n

$$F(n) = T c_n \tag{3.48}$$

Therefore we have

$$f(x) = \sum_{n=-\infty}^{\infty} \frac{F(n)}{T} \exp\left(\mathrm{i}\frac{2\pi n x}{T}\right) \tag{3.49}$$

$$F(n) = \int_{-T/2}^{T/2} f(x) \exp\left(-\mathrm{i}\frac{2\pi n x}{T}\right) \mathrm{d}x. \tag{3.50}$$

The Fourier coefficients of c_n have values at discrete points v_n with the period $1/T$, that is,

$$v_n = n/T. \tag{3.51}$$

Next, in the case when $T \to \infty$, c_n becomes a real continuous value, and therefore, we should employ a real variable v in spite of v_n. Eq. (3.50) is rewritten as

$$F(v) = \int_{-\infty}^{\infty} f(x) \exp(-\mathrm{i}2\pi v x) \mathrm{d}x. \tag{3.52}$$

Similarly, Eq. (3.49) is given by

$$f(x) = \int_{-\infty}^{\infty} F(v) \exp(\mathrm{i}2\pi v x) \mathrm{d}v. \tag{3.53}$$

The function $F(v)$ is called Fourier transform of the function $f(x)$, and $f(x)$ is called inverse Fourier transform of $F(v)$. $f(x)$ and $F(v)$ are called a Fourier transform pair and its relation is described by

$$f(x) \Leftrightarrow F(v). \tag{3.54}$$

Fourier transform is considered as a mapping operator

$$\mathscr{F}[f(x)] = F(v) \tag{3.55}$$

and its inverse mapping is $\mathscr{F}^{-1}$, then we have

$$\mathscr{F}^{-1}[F(v)] = f(x) \tag{3.56}$$

and

$$\mathscr{F}^{-1}\mathscr{F}[f(x)] = \mathscr{F}\mathscr{F}^{-1}[f(x)] = f(x). \tag{3.57}$$

3.5   SOME REPRESENTATIONS OF FOURIER TRANSFORM

The Fourier transform is defined using the angular frequency $\omega = 2\pi v$,

$$F(\omega) = \int_{-\infty}^{\infty} f(x) \exp(-\mathrm{i}\omega x) \mathrm{d}x. \tag{3.58}$$

Its inverse transform is given by

$$f(x) = \frac{1}{2\pi} \int_{-\infty}^{\infty} F(\omega) \exp(i\omega x)\,d\omega. \tag{3.59}$$

In the same manner, the following definition is also possible with opposite sign of imaginary unit,

$$F(\omega) = \int_{-\infty}^{\infty} f(x) \exp(i\omega x)\,dx \tag{3.60}$$

$$f(x) = \frac{1}{2\pi} \int_{-\infty}^{\infty} F(\omega) \exp(-i\omega x)\,d\omega. \tag{3.61}$$

In mathematics, a Fourier transform pair is defined as

$$F(\omega) = \frac{1}{\sqrt{2\pi}} \int_{-\infty}^{\infty} f(x) \exp(-i\omega x)\,dx \tag{3.62}$$

$$f(x) = \frac{1}{\sqrt{2\pi}} \int_{-\infty}^{\infty} F(\omega) \exp(i\omega x)\,d\omega. \tag{3.63}$$

All the definitions are equivalently valid. Different authors use different definitions according to the area of study. This can be confusing. Therefore make a note of the definition used.

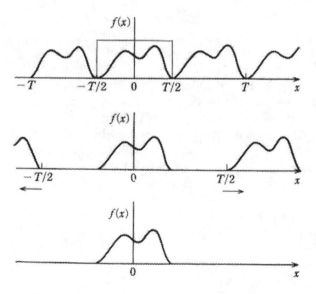

Figure 3.6 Extension of the Fourier series to the Fourier transform.

3.6 PROPERTIES OF THE FOURIER TRANSFORM

Consider the general properties of Fourier transform.

1. Linearity
    The Fourier transform is a linear transform, and therefore, the principle of superposition is valid.

    $$\mathscr{F}[a_1 f_1(x) + a_2 f_2(x)] = a_1 \mathscr{F}[f_1(x)] + a_2 \mathscr{F}[f_2(x)], \qquad (3.64)$$

    where a_1 and a_2 are constants.

2. Symmetricity (oddness and evenness)
    The symmetry properties play important roles in the Fourier transform. For example, consider a real function. From Eqs. (3.26) and (3.27), we can decompose a function $f(x)$ into an odd function $f_o(x)$ and an even function $f_e(x)$,

    $$
    \begin{aligned}
    F(\nu) &= \int_{-\infty}^{\infty} (f_e(x) + f_o(x)) \exp(-i 2\pi \nu x) \mathrm{d}x \\
    &= 2\left[\int_0^{\infty} f_e(x) \cos(2\pi \nu x) \mathrm{d}x - i \int_0^{\infty} f_o(x) \sin(2\pi \nu x) \mathrm{d}x \right] \\
    &= F_e(\nu) + F_o(\nu),
    \end{aligned}
    \qquad (3.65)
    $$

    where

    $$F_e(\nu) = 2 \int_0^{\infty} f_e(x) \cos(2\pi \nu x) \mathrm{d}x \qquad (3.66)$$

    $$F_o(\nu) = -2 \int_0^{\infty} f_o(x) \sin(2\pi \nu x) \mathrm{d}x. \qquad (3.67)$$

    Equations (3.66) and (3.67) correspond to the coefficients a_n and b_n of Fourier series and called Fourier cosine transform and Fourier sine transform, respectively. This means that the Fourier transform of a real even function is a real function and given by Fourier cosine transform, and the Fourier transform of a real odd function is imaginary and is given by Fourier sine transform.
    As coefficients c_n in Fourier series are called the spectrum, $F(\nu)$ is also called spectrum, and ν is called frequency. It should be noted that ν is not a discrete integer but a continuous real number because $f(x)$ is a non-periodic function. Table 3.2 shows symmetricity properties of Fourier transform, where Hermitian means the real part of a function is even and its imaginary part is odd, and anti-Hermitian means the real part of a function is odd and its imaginary part is even.

3. Similarity
    Suppose $f(x)$ and $F(\nu)$ are a Fourier transform pair, we have

    $$
    \begin{aligned}
    \mathscr{F}\left[f\left(\frac{x}{a}\right) \right] &= \int_{-\infty}^{\infty} f\left(\frac{x}{a}\right) \exp(-i 2\pi \nu x) \mathrm{d}x \\
    &= |a| \int_{-\infty}^{\infty} f(x) \exp(-i 2\pi \nu a x) \mathrm{d}x \\
    &= |a| F(a\nu),
    \end{aligned}
    \qquad (3.68)
    $$

Table 3.2

Symmetricity Properties of Fourier Transform

$f(x)$	$F(v)$
complex, asymmetrical	complex, asymmetrical
Hermitian	real, asymmetrical
antiHermitian	imaginary, asymmetrical
complex, even	complex, even
complex, odd	complex, odd
real, asymmetrical	Hermitian
real, even	real, even
real, odd	imaginary, odd
imaginary, asymmetrical	antiHermitian
imaginary, even	imaginary, even
imaginary, odd	real, odd

where a is a non-zero real number. This means that a signal with a narrower extent has a spectrum with a wider extent and the reverse is also true, as shown in Fig. 3.7.

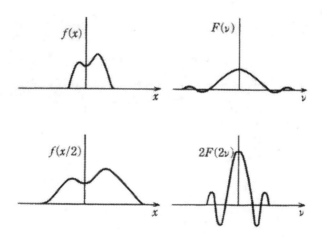

Figure 3.7 Similarity theorem of Fourier transform.

4. Shift theorem

Consider a signal $f(x)$ shifted by an amount a,

$$
\begin{aligned}
\mathscr{F}[f(x-a)] &= \int_{-\infty}^{\infty} f(x-a)\exp(-i2\pi vx)dx \\
&= \int_{-\infty}^{\infty} f(x)\exp[-i2\pi v(x+a)]dx \\
&= \exp(-i2\pi av)F(v).
\end{aligned}
\tag{3.69}
$$

Suppose a signal is real, we have

$$
\begin{aligned}
F(v) &= \int_{-\infty}^{\infty} f(x)\exp(-i2\pi vx)dx \\
&= \int_{-\infty}^{\infty} f(x)\cos(2\pi vx)dx - i\int_{-\infty}^{\infty} f(x)\sin(2\pi vx)dx
\end{aligned}
\tag{3.70}
$$

and therefore,

$$
\text{Re}[F(v)] = \int_{-\infty}^{\infty} f(x)\cos(2\pi vx)dx
\tag{3.71}
$$

$$
\text{Im}[F(v)] = -\int_{-\infty}^{\infty} f(x)\sin(2\pi vx)dx.
\tag{3.72}
$$

Table 3.3 shows the properties of the Fourier transform.

Table 3.3
Properties of Fourier Transform

$f(x)$	$F(v)$
$f(\pm x)$	$F(\pm v)$
$f^*(\pm x)$	$F^*(\mp v)$
$F(\pm x)$	$f(\mp v)$
$F^*(\pm x)$	$f^*(\mp v)$
$f(ax)$	$\dfrac{1}{|a|}F\left(\dfrac{v}{a}\right)$
$f(x\pm a)$	$F(v)\exp(\pm i2\pi av)$
$f(x)\cos(2\pi ax)$	$\dfrac{1}{2}[F(v+a)+F(v-a)]$
$f(x)\sin(2\pi ax)$	$\dfrac{i}{2}[F(v+a)-F(v-a)]$
$\dfrac{d^n f}{dx^n}$	$(i2\pi v)^n F(v)$
$(-i2\pi x)^n f(x)$	$\dfrac{d^n F(v)}{dv^n}$

3.7   DELTA FUNCTION

To consider applications of Fourier transform, we introduce a function corresponding to an impulse. Many types of impulse functions can be discussed. For example, a gaussian function, which is integral is 1 and width is infinitely small, as described by

$$\delta(x) = \lim_{w \to 0} \frac{1}{w} \exp\left(-\pi \frac{x^2}{w^2}\right). \tag{3.73}$$

This "function" is called generalized function or distribution, which is not a function by a strict definition. This $\delta(x)$ is interpreted as a kind of function that approaches to infinity at $x = 0$ by decreasing its width, but its integral value is unity. This is a very useful function to describe a point light source, a point object and a point image in this book and is call the Dirac δ function.

The Heaviside step function is defined by

$$H(x) = \begin{cases} 1 : x > 0 \\ 0 : x \leq 0. \end{cases} \tag{3.74}$$

This function is undifferentiable at $x = 0$ but its formal differential has a property of the delta function $\delta(x)$, so we have

$$H'(x) = \delta(x) \tag{3.75}$$

For a mathematically well-behaved function $f(x)$, its integration by parts is given by

$$\int_{-\infty}^{\infty} f(x)\delta(x)dx = \int_{-\infty}^{\infty} f(x)H'(x)dx$$

$$= \left[f(x)H(x)\right]_{-\infty}^{\infty} - \int_{-\infty}^{\infty} f'(x)H(x)dx$$

$$= -\int_{0}^{\infty} f'(x)dx = -[f(x)]_{0}^{\infty} = f(0). \tag{3.76}$$

This means the delta function $\delta(x)$ has the property

$$\int_{-\infty}^{\infty} f(x)\delta(x)dx = f(0). \tag{3.77}$$

Conversely, the delta function $\delta(x)$ can be defined by Eq. (3.77).

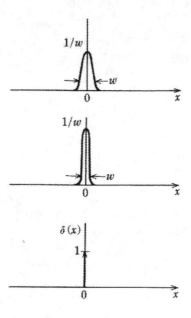

Figure 3.8   Concept of a delta function $\delta(x)$.

The delta function $\delta(x)$ is a non-periodic function but an array of delta functions with a period T given by

$$\delta_T(x) = \sum_{n=-\infty}^{\infty} \delta(x - nT) \tag{3.78}$$

is a periodic function. This means its Fourier series can be obtained. Its coefficients are given by

$$a_0 = \frac{2}{T} \int_{-T/2}^{T/2} \delta_T(x)\mathrm{d}x = \frac{2}{T} \tag{3.79}$$

$$a_n = \frac{2}{T} \int_{-T/2}^{T/2} \delta_T(x) \cos\left(\frac{2\pi nx}{T}\right)\mathrm{d}x = \frac{2}{T} \tag{3.80}$$

$$b_n = \frac{2}{T} \int_{-T/2}^{T/2} \delta_T(x) \sin\left(\frac{2\pi nx}{T}\right)\mathrm{d}x = 0. \tag{3.81}$$

Finally, the Fourier series of the function $\delta_T(x)$ is obtained as

$$\delta_T(x) = \frac{1}{T} + \frac{2}{T} \sum_{n=1}^{\infty} \cos\left(\frac{2\pi nx}{T}\right). \tag{3.82}$$

Since its Fourier coefficients $2/T$ are independent of n, this Fourier series does not converge. Similarly, the delta function is formally described by $\delta(x)$, which is divergent at $x = 0$, this Fourier series is formally written as $\delta_T(x)$.

Table 3.4

Fourier Transform Pairs Including $\delta(x)$

$f(x)$	$F(v)$
$\delta(x)$	1
1	$\delta(v)$
$\delta(x \pm a)$	$\exp(\pm i2\pi av)$
$\cos(2\pi ax)$	$\frac{1}{2}[\delta(v+a)+\delta(v-a)]$
$\sin(2\pi ax)$	$\frac{i}{2}[\delta(v+a)-\delta(v-a)]$

Consider the Fourier transform of the delta function $\delta(x)$.

$$\mathscr{F}[\delta(x)] = \int_{-\infty}^{\infty} \delta(t)\exp(-i2\pi vx)dx = \exp(0) = 1. \qquad (3.83)$$

This means that Fourier transform of the delta function is identically unity. By using this property, we have Fourier transform of $\exp(i2\pi ax)$,

$$\mathscr{F}[\exp(i2\pi ax)] = \int_{-\infty}^{\infty} \exp(i2\pi ax)\cdot\exp(-i2\pi vx)dx$$

$$= \int_{-\infty}^{\infty} 1\cdot\exp[-i2\pi x(v-a)]dx$$

$$= \delta(v-a). \qquad (3.84)$$

Similarly, we have

$$\mathscr{F}[\cos(2\pi ax)] = \mathscr{F}\left[\frac{1}{2}\exp(i2\pi ax) + \frac{1}{2}\exp(-i2\pi ax)\right]$$

$$= \frac{1}{2}[\delta(v-a)+\delta(v+a)] \qquad (3.85)$$

$$\mathscr{F}[\sin(2\pi ax)] = \mathscr{F}\left[\frac{1}{2i}\exp(i2\pi ax) - \frac{1}{2i}\exp(-i2\pi ax)\right]$$

$$= -i\frac{1}{2}[\delta(v-a)-\delta(v+a)]. \qquad (3.86)$$

The Fourier transform pairs related to the delta function is shown in Table 3.4.

3.8 CONVOLUTION INTEGRAL AND CORRELATION FUNCTION

Convolution and correlation are very important concepts in some scientific and engineering areas, such as linear response systems in Chapter 4. Consider two functions, $f_1(x)$ and $f_2(x)$, such that

$$f_1 * f_2(x) = \int_{-\infty}^{\infty} f_1(x')f_2(x-x')dx'. \qquad (3.87)$$

This is called convolution or convolution integral. The convolution is the overlapped area of functions $f_1(x')$ and $f_2(x-x')$ with a shifting parameter of x. It should be noted that the function $f_2(x-x')$ is a function turned over right to left and shift x. The convolution operation is illustrated in Fig. 3.9.

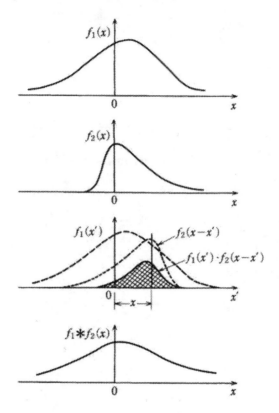

Figure 3.9   Convolution integral of $\int_{-\infty}^{\infty} f_1(x')f_2(x-x')dx'$.

Consider the Fourier transform of a convolution of $f_1(x)$ and $f_2(x)$, whose Fourier transforms are $F_1(v)$ and $F_2(v)$, respectively.

$$
\begin{aligned}
\mathscr{F}[f_1 * f_2] &= \iint_{-\infty}^{\infty} \left[f_1(x')f_2(x-x')dx' \cdot \exp(-i2\pi vx) \right] dx \\
&= \int_{-\infty}^{\infty} f_1(x') \left[\int_{-\infty}^{\infty} f_2(x-x')\exp(-i2\pi vx)dx \right] dx' \\
&= \int_{-\infty}^{\infty} f_1(x')F_2(v)\exp(-i2\pi vx')dx' \\
&= F_1(v)F_2(v).
\end{aligned}
\tag{3.88}
$$

The Fourier transform of a convolution integral gives us the product of Fourier transforms of each function. Conversely, the Fourier transform of the product of two functions gives us the convolution integral of the two functions, as described by

$$f_1 * f_2(x) \Leftrightarrow F_1(v) \cdot F_2(v). \tag{3.89}$$

This is called the convolution theorem.

On the other hand, the correlation function of two functions $f_1(x)$ and $f_2(x)$ is defined by

$$f_1 \star f_2^*(x) = \int_{-\infty}^{\infty} f_1(x') f_2^*(x' - x) dx', \tag{3.90}$$

where $f^*(x)$ denotes the complex conjugate of $f(x)$. As shown in Fig. 3.10, in the correlation, one of the functions is not turned over right to left, but its complex conjugate is used, dissimilar to the convolution.

The Fourier transform of the correlation is given by

$$\mathscr{F}[f_1 \star f_2^*(x)]$$
$$= \iint_{-\infty}^{\infty} f_1(x') f_2^*(x' - x) dx' \cdot \exp(-i2\pi vx) dx$$
$$= F_1(v) F_2^*(v) \tag{3.91}$$

and we have

$$f_1 \star f_2^*(x) \Leftrightarrow F_1(v) \cdot F_2^*(v). \tag{3.92}$$

This is called the correlation.

In general, $f_1(x)$ and $f_2(x)$ are different, and therefore the correlation function is called the cross-correlation function, but in the case of two functions being identical, this is called the autocorrelation function. Obviously we have from Eq. (3.91),

$$f \star f^*(x) \Leftrightarrow |F(v)|^2, \tag{3.93}$$

which gives us the square of the absolute of its Fourier transform. Rewriting Eq. (3.93), we have

$$\iint_{-\infty}^{\infty} f(x') f^*(x' - x) dx' = \int_{-\infty}^{\infty} |F(v)|^2 \exp(i2\pi vx) dv \tag{3.94}$$

Consider the case $x = 0$ in Eq. (3.94), we have

$$\int_{-\infty}^{\infty} |f(x)|^2 dx = \int_{-\infty}^{\infty} |F(v)|^2 dv. \tag{3.95}$$

This is called the Parseval theorem. Analogies to the discussion of Eq. (1.43), if $f(x)$ is the amplitude of waves, $|f(x)|^2$ gives the energy of the wave. Therefore, $|F(v)|^2$ corresponds to the energy of the Fourier spectrum, that is, the energy of the spectrum, called the power spectrum, or the energy spectrum. Namely, the Parseval theorem means that the energy in the real space is equal to the energy in the spectrum space. The energy conservation rule is valid between the real space and the spectrum space.

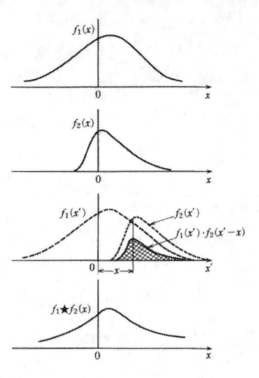

Figure 3.10   Correlation of $\int_{-\infty}^{\infty} f_1(x') f_2(x'-x) dx'$ in the case when $f_1(x)$ and $f_2(x)$ are real.

3.9   SOME FUNCTIONS AND THEIR FOURIER TRANSFORMS

In this section, some functions used in future discussions are defined and their Fourier transforms are described.

1. Rectangular function

$$\text{rect}(x) = \begin{cases} 1 : |x| \leq \frac{1}{2} \\ 0 : \text{otherwise.} \end{cases} \tag{3.96}$$

This function has a rectangular shape, as shown in Fig. 3.11(a). This rectangular function rect(x) has the operation that extracts apart from $-1/2$ to $1/2$ along to the x-axis out of a function $f(x)$. That is, $f(x) \cdot \text{rect}(x)$ gives the range of $f(x)$ from $f(-1/2)$ to $f(1/2)$, as shown in Fig. 3.11(b). A function with this property is called a window function or a gate function. Such a function is used to describe sometimes a slit operation and the location of an object in optical system analysis.

2. Sinc function

$$\text{sinc}(x) = \frac{\sin(\pi x)}{\pi x} \tag{3.97}$$

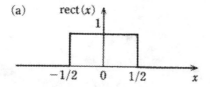

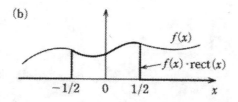

Figure 3.11 (a) The rectangular function rect(x) and (b) its operation as a window function.

3. Triangular function

$$\Lambda(x) = \begin{cases} 1 - |x| : |x| \leq 1 \\ 0 : \text{otherwise} \end{cases} \tag{3.98}$$

4. Comb function

$$\text{comb}(x) = \sum_{n=-\infty}^{\infty} \delta(x - n) \tag{3.99}$$

5. Sign function (Signum function)

$$\text{sgn} = \begin{cases} 1 : x > 0 \\ 0 : x = 0 \\ -1 : x < 0 \end{cases} \tag{3.100}$$

6. Circular function

$$\text{circ}(r) = \begin{cases} 1 : r = \sqrt{x^2 + y^2} \leq 1 \\ 0 : \text{otherwise} \end{cases} \tag{3.101}$$

7. Gaussian function

$$\text{gauss}(r) = \exp(-\pi r^2) \tag{3.102}$$

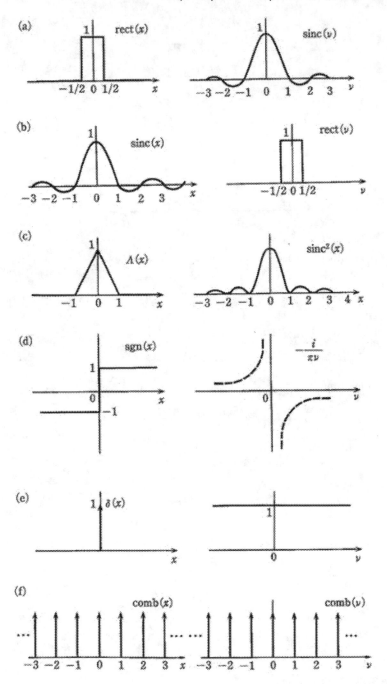

Figure 3.12 Typical functions and their Fourier transforms(1).

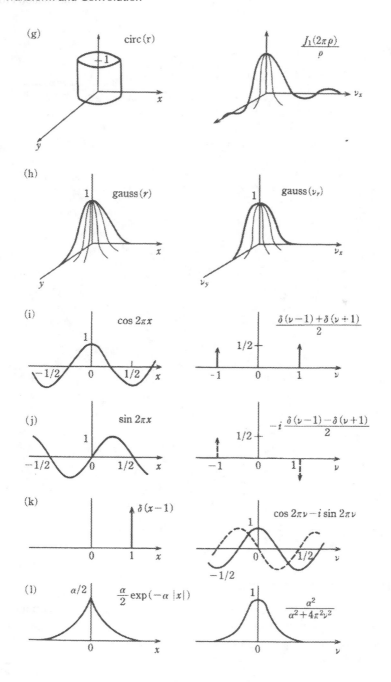

Figure 3.13 Typical functions and their Fourier transforms(2).

Table 3.5
Fourier Transform Pairs(1)

$f(x) = \int_{-\infty}^{\infty} F(v)\exp(\mathrm{i}2\pi vx)\mathrm{d}v$	$F(v) = \int_{-\infty}^{\infty} f(x)\exp(-\mathrm{i}2\pi vx)\mathrm{d}x\mathrm{d}y$
1	$\delta(v)$
$\delta(x)$	1
$\mathrm{gauss}(x) = \exp(-\pi x^2)$	$\mathrm{gauss}(v) = \exp(-\pi v^2)$
$\cos(2\pi v_0 x)$	$\dfrac{1}{2}[\delta(v - v_0) + \delta(v + v_0)]$
$\sin(2\pi v_0 x)$	$-\dfrac{\mathrm{i}}{2}[\delta(v - v_0) - \delta(v + v_0)]$
$\mathrm{rect}(x)$	$\mathrm{sinc}(v)$
$\Lambda(x)$	$\mathrm{sinc}^2(v)$
$\exp(-|x|)$	$\dfrac{2}{1 + (2\pi v)^2}$
$J_0(2\pi x)$	$\dfrac{\mathrm{rect}(v/2)}{\pi(1 - v^2)^{1/2}}$
$\dfrac{J_1(2\pi x)}{2x}$	$(1 - v^2)^{1/2}\mathrm{rect}(v/2)$
$\mathrm{sgn}(x)$	$-\dfrac{\mathrm{i}}{\pi v}$
$H(x)$	$\dfrac{1}{2}\delta(v) - \dfrac{\mathrm{i}}{2\pi}$
$\exp(\mathrm{i}\pi x^2)$	$\exp\left[\mathrm{i}\left(\dfrac{\pi}{4}\right)\right]\exp(-\mathrm{i}\pi v^2)$
$\mathrm{comb}(ax)$	$\dfrac{1}{|a|}\mathrm{comb}(\dfrac{v}{a})$
x^k	$\left(\dfrac{\mathrm{i}}{2\pi}\right)^k\delta^{(k)}(v)$
$\mathrm{sech}(\pi x)$	$\mathrm{sech}(\pi v)$
$\dfrac{1}{\sqrt{|x|}}$	$\dfrac{1}{\sqrt{|v|}}$

The shapes and their Fourier transforms are shown in Figs. 3.12 and 3.13, and Tables 3.5 and 3.6.

3.10   SAMPLING THEORY

Until now, we have discussed Fourier transform of a continuous function of x, but in the case of a discrete array of data for alternatively sampling discrete points of a continuous function, we have to evaluate a numerical Fourier transform by a computer. The integral

$$\int_{-\infty}^{\infty} f(x)\delta(x - x_0)\mathrm{d}x = f(x_0) \tag{3.103}$$

is very useful to describe sampling a continuous function $f(x)$ at a point $x = x_0$. As shown in Fig. 3.14, in sampling N points at $x = nT$ ($n = 0, \pm 1, \pm 2, \ldots$) with a

Table 3.6
Fourier Transform Pairs(2)

$f(x,y) = \int_{-\infty}^{\infty} F(v_x, v_y)$ $\exp[i2\pi(v_x x + v_y y)]dv_x dv_y$	$F(v_x, v_y) = \int_{-\infty}^{\infty} f(x,y)$ $\exp[-i2\pi(v_x x + v_y y)]dx dy$
1	$\delta(v_x, v_y)$
$\delta(x,y)$	1
$\text{rect}(x)\text{rect}(y)$	$\text{sinc}(v_x)\text{sinc}(v_y)$
$\Lambda(x)\Lambda(y)$	$\text{sinc}^2(v_x)\text{sinc}^2(v_y)$
$\text{gauss}(r) = \exp(-\pi r^2) = \exp[-\pi(x^2 + y^2)]$	$\text{gauss}(\rho) = \exp(-\pi\rho^2) = \exp[-\pi(v_x^2 + v_y^2)]$
$\text{circ}(r)$	$\dfrac{J_1(2\pi\rho)}{\rho}$
$\delta(r-a)$	$2\pi a J_0(2\pi a\rho)$
$\dfrac{1}{r}$	$\dfrac{1}{\rho}$
$\exp(i\pi r^2) = \exp[i\pi(x^2 + y^2)]$	$i\exp(-i\pi\rho^2) = i\exp[-i\pi(v_x^2 + v_y^2)]$

where $r = \sqrt{x^2 + y^2}$ and $\rho = \sqrt{v_x^2 + v_y^2}$.

sampling period of T, we have sampled data array

$$f_s(x) = f(x) \cdot \text{comb}\left(\frac{x}{T}\right). \tag{3.104}$$

It should be noted that

$$\text{comb}\left(\frac{x}{T}\right) = T \sum_{n=-\infty}^{\infty} \delta(x - nT). \tag{3.105}$$

From Fig. 3.12(f), we have

$$\mathscr{F}[\text{comb}(x)] = \text{comb}(v) \tag{3.106}$$

and therefore

$$\mathscr{F}\left[\text{comb}\left(\frac{x}{T}\right)\right] = T\text{comb}(Tv) \tag{3.107}$$

by using the similarity rule in Fourier transform given by Eq. (3.68). By Fourier transforming the both sides of Eq. (3.104), we have

$$F_s(v) = F(v) * \mathscr{F}\left[\text{comb}\left(\frac{x}{T}\right)\right]. \tag{3.108}$$

Equation (3.108) is rewritten by

$$F_s(v) = F(v) * T\text{comb}(Tv). \tag{3.109}$$

Since

$$T\text{comb}(Tv) = \sum_{n=-\infty}^{\infty} \delta\left(v - \frac{n}{T}\right), \tag{3.110}$$

finally we have the sampled Fourier spectrum of Eq. (3.109)

$$F_s(v) = \sum_{n=-\infty}^{\infty} F\left(v - \frac{n}{T}\right). \tag{3.111}$$

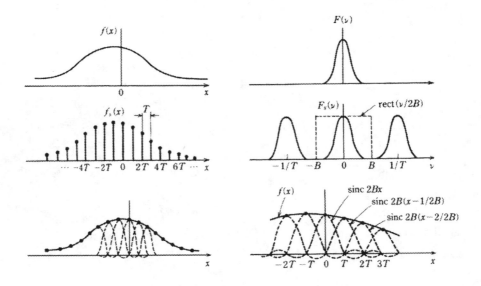

Figure 3.14   Sampling theory.

Figure 3.14 shows the sampling process and a sampled function $f_s(x)$ and its Fourier transform $F_s(v)$. Equation (3.111) means that the sampled Fourier spectrum $F_s(v)$ is an array of the spectrum $F(v)$ of a continuous function $f(x)$ with a period of $1/T$.

To recover the original continuous function $f(x)$ from its sampled function $f_s(x)$, we should take $F(v)$ out of a periodic function $F_s(v)$ in the Fourier domain. To do so, it is necessary that each spectrum component in $F_s(v)$ should be located with enough separation. If the spectrum components are overlapped, the original spectrum $F(v)$ is not recovered correctly. Consider the spectrum width of $F(v)$, called the bandwidth is $2B$. The condition of non-overlapping is given by

$$\frac{1}{T} \geq 2B \tag{3.112}$$

because the period of the spectrum $F_s(v)$ is $1/T$. The signal with a limited spectrum range is called a band-limited signal.

To obtain $F(v)$ from $F_s(v)$, we employ a window function $\text{rect}(v/2B)$ with a width of $2B$. The original function is recovered as

$$f(x) = \mathscr{F}^{-1}\left[F_s(v)\text{rect}\left(\frac{v}{2B}\right)\right]$$

$$= \mathscr{F}^{-1}[F_s(v)] * \mathscr{F}^{-1}\left[\text{rect}\left(\frac{v}{2B}\right)\right]. \tag{3.113}$$

From Eq. (3.104), we have

$$\mathscr{F}^{-1}[F_s(v)] = f_s(x) = T \sum_{n=-\infty}^{\infty} f(nT)\delta(x - nT) \tag{3.114}$$

and from Table 3.5, the Fourier transform of the rect function is given by

$$\mathscr{F}^{-1}\left[\text{rect}\left(\frac{v}{2B}\right)\right] = 2B\text{sinc}(2Bx) \tag{3.115}$$

Finally, Eq. (3.113) is rewritten by

$$f(x) = 2BT \sum_{n=-\infty}^{\infty} f(nT)\text{sinc}[2B(x - nT)]. \tag{3.116}$$

By using a critical sampling period $T = 1/2B$, the original band-limited function $f(x)$ is given by

$$f(x) = \sum_{n=-\infty}^{\infty} f\left(\frac{n}{2B}\right)\text{sinc}\left[2B\left(x - \frac{n}{2B}\right)\right]. \tag{3.117}$$

This means that a continuous function $f(x)$ can be recovered by using only the array of sampling data $f(n/2B)$. This is called the sampling theorem. It should be noted that the sampling theorem is valid when the original function is band-limited (bandwidth 2B) and the sampling period is less than $1/2B$.

In the case when $T \geq 1/2B$, the spectrum bands are overlapped and higher components of spectrum are folded in as shown in Fig.3.15, and then errors are generated in spectrum. This type of error is called an aliasing error.

To avoid the aliasing error, low-pass filtering of the signal before sampling is necessary.

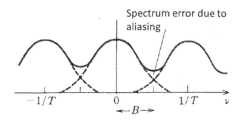

Spectrum error due to aliasing

$-1/T$ 0 $1/T$ v

$\leftarrow B \rightarrow$

Figure 3.15 Aliasing error.

PROBLEMS

1. Determine the Fourier series of

$$f(x) = \cos(\pi x)$$

   within $[-1/2, 1/2]$.
2. Determine the Fourier series of

$$f(x) = x^2$$

   within $[-1/2, 1/2]$.
3. Determine the Fourier transforms

   a.

$$f(x) = \exp(i2\pi\alpha x^2)$$

   b.

$$f(x) = \exp(-2\pi\alpha|x|), \qquad (\alpha > 0)$$

   c.

$$f(x) = \exp(-\alpha|x|)\cos(2\pi\nu_0 x)$$

4. Draw graphs of convolution and correlation functions of functions $f_1(x)$ and $f_2(x)$ shown in Fig. 3.16.
5. Determine the convolution integral

$$\exp(-\alpha x^2) * \exp(-\beta x^2), \qquad (\alpha, \beta > 0)$$

   with and without the Fourier transform.
6. Prove the following equations.

   a.

$$\delta(ax) = \frac{1}{|a|}\delta(x).$$

   b.

$$\mathscr{F}[\text{comb}(x)] = \text{comb}(\nu_x)$$

   c.

$$\text{comb}(ax) = \frac{1}{|a|}\sum \delta(x - n/a) \qquad (3.118)$$

7. Determine the Parseval theorem of the following equations

   a.

$$f(x) = \exp(-|x|)$$

   b.

$$f(x) = \text{rect}(x)$$

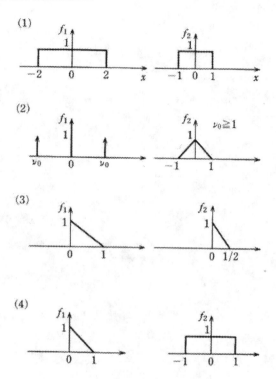

Figure 3.16 Functions $f_1(x)$ and $f_2(x)$ for Problems 4.

and show the following integrals from former results

a.

$$\int_{-\infty}^{\infty} \frac{dv}{(1+v^2)^2} = \frac{\pi}{2}$$

b.

$$\int_0^{\infty} \frac{\sin^2 \pi v}{\pi^2 v^2} dv = \frac{1}{2}.$$

BIBLIOGRAPHY

Papoulis, A. 1968. *Systems and Transforms with Applications in Optics*. McGraw-Hill, New York.

Bracewell, R. N. 1986. *The Fourier Transform and Its Applications*. McGraw-Hill, New York.

4 Linear System

In many cases, when we analyze scientific and engineering systems as well as optical systems, we use the method in which characteristics of a system are estimated or evaluated by the relation between its input and output. This approach enables us to analyze the general characteristics of the system without the knowledge of the inner structure and physical properties of the system. This method looks simple but very valuable in many cases when we determine the general properties of the system or understand the characteristics of the system by analogy with the other systems.

In this chapter, the linear shift-invariant system is mainly discussed and its fundamental mathematical concepts and mathematical treatments are described.

4.1 SYSTEM AND OPERATOR

When we analyze a complicated phenomenon under certain circumstances, we can use an approach to find general properties of the phenomenon only from the event that causes the phenomenon and the event that results from the phenomenon. Consider an arbitrary system with the input $f(x)$ and output $g(x)$, as shown in Fig. 4.1. We describe this system as

$$\mathscr{S}[f(x)] = g(x). \tag{4.1}$$

This means a function $f(x)$ is transformed or mapped to a function $g(x)$ by $\mathscr{S}$, which is called an operator and $\mathscr{S}$ is often called a system.

The concept of the operator is very general. For example, the wave equation of Eq. (1.9)

$$\frac{\partial^2 u}{\partial x^2} + \frac{\partial^2 u}{\partial y^2} + \frac{\partial^2 u}{\partial z^2} = \frac{1}{v^2}\frac{\partial^2 u}{\partial t^2} \tag{4.2}$$

is rewritten as

$$\nabla^2 u = \frac{1}{v^2}\frac{\partial^2 u}{\partial t^2} \tag{4.3}$$

by using the Laplacian operator

$$\nabla^2 = \frac{\partial^2}{\partial x^2} + \frac{\partial^2}{\partial y^2} + \frac{\partial^2}{\partial z^2}. \tag{4.4}$$

This means that the wave equation is a system, which maps $u(x,y,z)$ to $\frac{1}{v^2}\frac{\partial^2 u}{\partial t^2}$ by the Laplacian operator.

As another example, the Fourier transform operator $\mathscr{F}$ has been introduced in Eq. (3.55). The Fourier transform operator maps a function $f(x)$ in real space into a function $F(v)$ in the Fourier space or the frequency space.

DOI: 10.1201/9781003121916-4

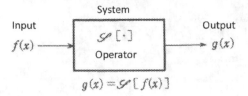

Figure 4.1   Linear system.

4.2   LINEAR SYSTEM AND SHIFT-INVARIANT SYSTEM

4.2.1   LINEAR SYSTEM

Consider a system described by an operator $\mathscr{L}$, and two inputs $f_1(x)$ and $f_2(x)$ and their outputs $g_1(x)$ and $g_2(x)$, respectively.

$$\mathscr{L}[f_1(x)] = g_1(x) \tag{4.5}$$

$$\mathscr{L}[f_2(x)] = g_2(x). \tag{4.6}$$

The system is called a linear system when it obeys the linearity conditions:

$$\begin{aligned}
\mathscr{L}[a_1 f_1(x) &+ a_2 f_2(x)] \\
&= \mathscr{L}[a_1 f_1(x)] + \mathscr{L}[a_2 f_2(x)] \\
&= a_1 \mathscr{L}[f_1(x)] + a_2 \mathscr{L}[f_2(x)] \\
&= a_1 g_1(x) + a_2 g_2(x),
\end{aligned} \tag{4.7}$$

where a_1 and a_2 are constants. From analogy to Eq. (1.47), the principle of super-position is valid in a linear system. That is, linearity is a synonym for the principle of superposition. The Laplace operator and the Fourier transform operator are linear operators and the principle of superposition is valid in wave phenomena and Fourier transforms.

4.2.2   SHIFT-INVARIANT SYSTEM

Consider again a system $\mathscr{S}$, its input $f(x)$ and the output $g(x)$,

$$\mathscr{S}[f(x)] = g(x). \tag{4.8}$$

If the input $f(x)$ is shifted by an amount of x_0 and therefore the output $g(x)$ is also shifted by the same amount of x_0 as shown in Fig. 4.2, then we have

$$\mathscr{S}[f(x - x_0)] = g(x - x_0). \tag{4.9}$$

This system is invariant to the shift of x_0 and this system is called a shift-invariant system. The shift-invariant property for the space coordinate of x means that the characteristic of the system $\mathscr{S}$ is not changed and the shape of the output is not changed, but the position is changed depending on the position of the input. The shift-invariant system in time is sometimes called a stationary system or a time-invariant system. The stationary system is stable in time.

The linear shift-invariant system is defined as

$$\mathscr{S}[a_1 f_1(x-x_1) + a_2 f_2(x-x_2)] = a_1 g_1(x-x_1) + a_1 g_1(x-x_1) \qquad (4.10)$$

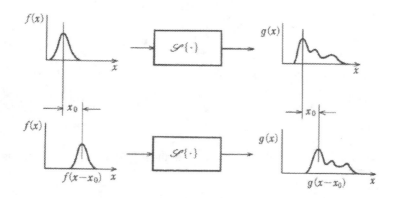

Figure 4.2   Shift-invariant system.

4.2.3   IMPULSE RESPONSE

Consider a linear shift-invariant system

$$\mathscr{T}[f(x)] = g(x) \qquad (4.11)$$

and the input is an impulse. The impulse is an idealized input of infinitesimal size, e.g., a point light source. The impulse is described by a delta function $\delta(x)$ and its response is given by

$$\mathscr{T}[\delta(x)] = h(x). \qquad (4.12)$$

Since the system is shift-invariant, we have

$$\mathscr{T}[\delta(x-x_0)] = h(x-x_0). \qquad (4.13)$$

The output of the impulse input is called the impulse response. It is amazing that all the characteristics of the linear shift-invariant system are described by the impulse response.

By using the principle of superposition of a linear system, we decompose the input $f(x)$ to a series of impulses

$$f_T(x) = \sum_{n=-\infty}^{\infty} f(x)\delta(x-nT). \qquad (4.14)$$

If the period T of impulses is decreased to the infinitesimal, we have, as shown in Fig. 4.3.

$$f(x) = \int_{-\infty}^{\infty} f(x')\delta(x-x')dx'. \tag{4.15}$$

Equation (4.15) is the same as the definition of the delta function $\delta(x)$ in Eq. (3.77). By substituting Eq. (4.15) to Eq. (4.11), we have

$$g(x) = \mathscr{T}\left[\int_{-\infty}^{\infty} f(x')\delta(x-x')dx'\right]$$
$$= \int_{-\infty}^{\infty} f(x')\mathscr{T}[\delta(x-x')]dx'. \tag{4.16}$$

By substituting Eq. (4.13) to Eq. (4.16),

$$g(x) = \int_{-\infty}^{\infty} f(x')h(x-x')dx'$$
$$= f * h(x). \tag{4.17}$$

The output of a linear shift-invariant system is the convolution of the input and the impulse response.

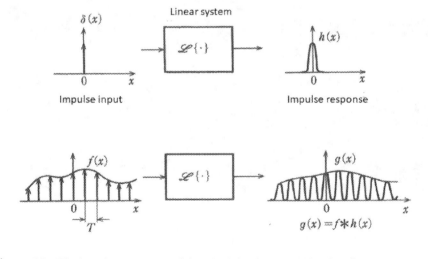

Figure 4.3 The impulse response and the principle of superposition in a linear system.

4.3 FREQUENCY RESPONSE FUNCTION

Consider the input $f(x)$ and output $g(x)$ of a linear shift-invariant system and its impulse response $h(x)$. Each Fourier transform is given by

$$F(v) = \int_{-\infty}^{\infty} f(x)\exp(-i2\pi vx)dx \tag{4.18}$$

$$G(v) = \int_{-\infty}^{\infty} g(x) \exp(-i2\pi v x) dx \qquad (4.19)$$

$$H(v) = \int_{-\infty}^{\infty} h(x) \exp(-i2\pi v x) dx. \qquad (4.20)$$

The output of the system is given by the convolution of the input and the impulse response of the system as shown in Eq. (4.16) and using the convolution theorem, we have

$$G(v) = F(v) \cdot H(v). \qquad (4.21)$$

Here $H(v)$ is called the frequency response of the linear shift-invariant system. Equation (4.21) means that the system changes the input spectrum $F(v)$ to the output spectrum $G(v)$ by $H(v)$ times. That is, the frequency response function gives the magnitude of the system response in each frequency v. Since the frequency response and the impulse response are a Fourier transform pair and therefore the information of the both is the same, all the characteristics of the linear shift-invariant system are described by the frequency response, as shown in Fig. 4.4.

4.4   EIGENFUNCTION AND EIGENVALUE

The impulse response can describe a linear shift-invariant system. Another approach to describe the system characteristics is discussed.

Consider the case when the system output $\psi(x;\xi)$ is the same shape as the input $\psi(x;\xi)$ with appropriate attenuation $H(\xi)$.

$$\mathcal{T}[\psi(x;\xi)] = H(\xi)\psi(x;\xi), \qquad (4.22)$$

where ξ is a complex constant. We do not prove the existence of the solution here, but a practical example will be shown later.

When the input and the output of the system are given by Eq.(4.22), $\psi(x;\xi)$ is called the eigenfunction of the operator $\mathcal{T}$ and $H(\xi)$ the eigenvalue of the eigenfunction $\psi(x;\xi)$.

Consider the case that $\exp(i2\pi\xi x)$ is an input and its output $g(x;\xi)$.

$$\mathcal{T}[\exp(i2\pi\xi x)] = g(x;\xi). \qquad (4.23)$$

Then the input is shifted by x',

$$\begin{aligned}
\mathcal{T}&\{\exp[i2\pi\xi(x-x')]\} \\
&= \mathcal{T}[\exp(i2\pi\xi x)]\exp(-i2\pi\xi x') \\
&= g(x;\xi)\exp(-i2\pi\xi x').
\end{aligned} \qquad (4.24)$$

Since the system is shift-invariant, we have

$$\mathcal{T}\{\exp[i2\pi\xi(x-x')]\} = g(x-x':\xi). \qquad (4.25)$$

Finally we have the relationship

$$g(x-x':\xi) = \exp(-i2\pi\xi x')g(x;\xi). \qquad (4.26)$$

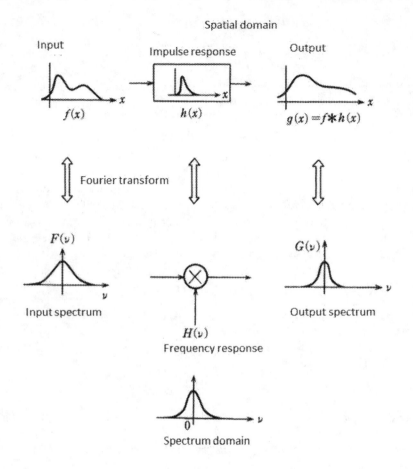

Figure 4.4   Relationship between impulse response and frequency response.

If we take

$$g(x:\xi) = H(\xi)\exp(i2\pi\xi x), \qquad (4.27)$$

it is obvious that Eq. (4.26) is valid, where $H(\xi)$ is a complex value. It should be noted that

$$\mathscr{T}[\exp(i2\pi\xi x)] = H(\xi)\exp(i2\pi\xi x). \qquad (4.28)$$

The function $\exp(i2\pi\xi x)$ is an eigenfunction of Eq. (4.22) and in the case $x = 0$, we have

$$H(\xi) = g(0;\xi). \qquad (4.29)$$

This means that the eigenvalue $H(\xi)$ is obtained by the output at the origin $x = 0$ for the system input $\exp(i2\pi\xi x)$. The use of the impulse response or frequency response has been discussed to describe the system characteristics. The response of a complex sinusoidal function $\exp(i2\pi\xi x)$ with the frequency ξ is $H(\xi)\exp(i2\pi\xi x)$, which is the output from the input $\exp(i2\pi\xi x)$ attenuated by $H(\xi)$. This means the eigenvalue $H(\xi)$ to the eigenfunction $\exp(i2\pi\xi x)$ is the frequency response itself. The frequency response is measured by the contrast of the sinusoidal output for the sinusoidal input with changing the frequency, as shown in Fig. 4.5.

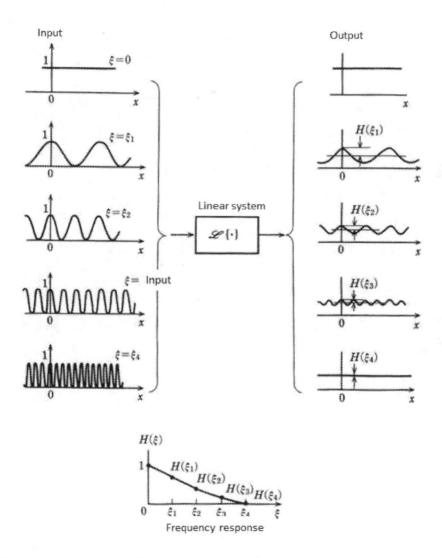

Figure 4.5 Eigenvalue $H(\xi)$ and frequency response.

PROBLEMS

1. Show three examples of the linear shift-invariant system in the field of science and engineering.

2. Determine the frequency response of the system whose shape of the output is not changed for the following input

$$f(x) = [1 + \cos(2\pi v_1 x)][1 + \cos(2\pi v_2 x)], \qquad (4.30)$$

where $v_1 > v_2$.

5 Discrete Fourier Transform and Fast Fourier Transform

The digital Fourier transform for a large number of sample points with a direct computation algorithm is typically causes an explosional increase in computational time. Especially in optical fields, we have to use 2-D or 3-D image data with a large number of sampling points. Their computational costs are large in some cases. Fortunately, an efficient algorithm, the Fast Fourier Transform (FFT), has been developed, which is widely used in optics, signal processing, physics, mathematics, and so on. In this chapter, we discuss the discrete Fourier transform, which is a digital version of Fourier transform and then window functions for performing discrete Fourier transform, and finally the algorithm of FFT and its programming.

5.1 DISCRETE FOURIER TRANSFORM

The sampling theorem shows that discrete data array $f_s(nT)$ is obtained by sampling a continuous signal $f(x)$ with an appropriate sampling period T. However, its Fourier spectrum $F_s(v)$ is a continuous periodic function. To do digital processing, the Fourier spectrum should be digitized. This procedure is shown in Fig. 5.1.

Let the Fourier spectrum of a continuous signal $f(x)$ be $F(v)$. The continuous signal is sampled by $\mathrm{comb}(x/T)$ with a period T,

$$f_s(x) = f(x) \times \mathrm{comb}\left(\frac{x}{T}\right). \tag{5.1}$$

Its spectrum is given by

$$\begin{aligned} F_s(v) &= \mathscr{F}[f_s(x)] \\ &= \int_{-\infty}^{\infty}\left[\sum_{n=-\infty}^{\infty} f(x)\delta(x-nT)\right]\exp(-\mathrm{i}2\pi vx)\mathrm{d}x \\ &= \sum_{n=-\infty}^{\infty} f(nT)\exp(-\mathrm{i}2\pi vnT). \end{aligned} \tag{5.2}$$

The spectrum $F_s(v)$ of the sampled function, as shown in Fig. 5.1(c), is periodical and continuous. From the sampling theorem, the spectrum $F_s(v)$ includes sufficient information in the range from $-1/2T$ to $1/2T$ under the condition $1/T \geq 2B$.

Because we can observe a limited number of discrete data, only sampled data at points $n = 0, 1, \cdots, N-1$ are used and the other points are cut off. Then the spectrum is written as

$$F_s(v) = \sum_{n=0}^{N-1} f(nT)\exp(-\mathrm{i}2\pi vnT), \tag{5.3}$$

DOI: 10.1201/9781003121916-5

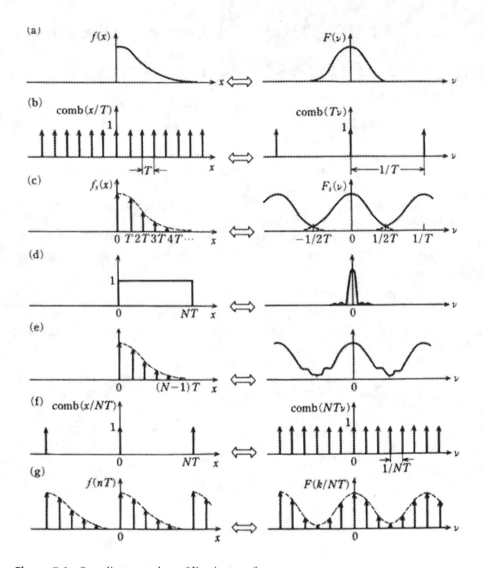

Figure 5.1 Sampling procedure of Fourier transform.

as shown in Fig. 5.1(e). Cutting off the outside points is the same as windowing of a rectangular window function within the range from 0 to NT. We call a signal with a limited frequency range the band-limited signal, a signal with a limited range of space is called space-limited signal.

Next, sample the space-limited spectrum $F_s(v)$. By applying the sampling theorem to the Fourier space, the sampling period is estimated as $1/NT$. Since the k-th sampling point is at $v = k/NT$ as shown in Fig. 5.1 (f), Eq. (5.3) is rewritten as

$$F\left(\frac{k}{NT}\right) = \sum_{n=0}^{N-1} f(nT) \exp\left(-\frac{i2\pi kn}{N}\right), \qquad (k = 0, 1, 2, \cdot, N-1). \qquad (5.4)$$

This is called the discrete Fourier transform (DFT) of $f(nT)$, which is a periodic signal with the period NT, as shown in Fig. 5.1(g). By using the orthogonality property,

$$\frac{1}{N} \sum_{n=0}^{N-1} \exp\left(-\frac{i2\pi kn}{N}\right) \exp\left(\frac{i2\pi mn}{N}\right) = \begin{cases} 1 : k-m \text{ is a multiple of } N \\ 0 : \text{otherwise}, \end{cases} \qquad (5.5)$$

$f(nT)$ is rewritten as

$$f(nT) = \frac{1}{N} \sum_{k=0}^{N-1} F\left(\frac{k}{NT}\right) \exp\left(\frac{i2\pi kn}{N}\right), \qquad (n = 0, 1, 2, \cdots, N-1). \qquad (5.6)$$

This is called the inverse discrete Fourier transform. For simplicity,

$$f(n) = f(nT) \qquad (5.7)$$

$$F(k) = F\left(\frac{k}{NT}\right) \qquad (5.8)$$

$$W = \exp\left(-\frac{i2\pi}{N}\right), \qquad (5.9)$$

we have the discrete Fourier transform and its inverse version

$$F(k) = \sum_{n=0}^{N-1} f(n)W^{kn} \qquad (5.10)$$

and

$$f(n) = \frac{1}{N} \sum_{k=0}^{N-1} F(k)W^{-kn}, \qquad (5.11)$$

respectively. Pairs of an original sinusoidal function and its sampling version: $f(x)$ and $f(n)$, and $F(v)$ and $F(k)$ are not always equal due to the cutting-off effect of sampling points. The sampling data array is only an approximation of a continuous function. On the other hand, it should be noted that the discrete Fourier transform pair of Eqs. (5.10) and (5.11) is rigorously related by the mathematical relationship.

5.2 WINDOW FUNCTIONS

The discrete Fourier transform is defined for periodic functions. In many applications, we can extract a limited part of the function and arrange it periodically to make a periodic function, even though the original function expands in relatively large area. This procedure is similar to observing a signal through a window.

A simple window is described as a rectangular function, where the value is 1 inside the window and otherwise 0. On both sides of the window, the values of the function have differences in some cases. This causes the discontinuity in the periodic function and some difficulties in Fourier transforming. In order to avoid this difficulty, weighted windows are discussed, in which both sides the value goes smoothly to zero.

The windowed function $f_w(x)$ is described as

$$f_w(x) = f(x) \cdot w(x), \tag{5.12}$$

where $f(x)$ and $w(x)$ denote the input function and the window function, respectively. The spectrum of the function is given by

$$F_w(v) = F(v) * W(v), \tag{5.13}$$

where $F_w(v)$, $F(v)$ and $W(v)$ denote the spectrum of $f_w(x)$, $f(x)$ and $w(x)$, respectively. Since the windowed function should be similar to the original function, $W(v)$ has a wide width and becomes zero rapidly and smoothly. The window functions should satisfy such contradictory conditions.

Typical window functions and their spectra are as follows and shown in Fig. 5.2.

1. Rectangular window function

$$w_1(x) = \text{rect}(x/T) \tag{5.14}$$

$$W_1(v) = T\text{sinc}(Tv) \tag{5.15}$$

2. Bartlett window function

$$w_2(x) = \Lambda\left(\frac{2x}{T}\right) \tag{5.16}$$

$$W_2(v) = \frac{T}{2}\text{sinc}^2\left(\frac{Tv}{2}\right) \tag{5.17}$$

3. Generalized Hamming window function[1]

$$w(x) = \begin{cases} \alpha + (1-\alpha)\cos\left(\frac{2\pi}{T}x\right) & : |x| \leq \frac{T}{2} \\ 0 & : \text{otherwise} \end{cases} \tag{5.18}$$

where $0 \leq \alpha \leq 1$. Especially, in the case when $\alpha = 0.5$, the window function is called Hanning window

$$w_3(x) = \begin{cases} \frac{1}{2}\left[1 + \cos\left(\frac{2\pi}{T}x\right)\right] & : |x| \leq \frac{T}{2} \\ 0 & : \text{otherwise} \end{cases} \tag{5.19}$$

$$W_3(v) = \frac{T}{2} \cdot \frac{\text{sinc}(Tv)}{1 - T^2v^2}. \tag{5.20}$$

---

[1] Hamming window and Hanning window originate from Richard W. Hamming and Julius von Hann, respectively.

In the case when $\alpha = 0.54$, the window function is simply called the Hamming window function,

$$w_4(x) = \begin{cases} 0.54 + 0.46\cos\left(\frac{2\pi}{T}x\right)\right] & : |x| \leq \frac{T}{2} \\ 0 & : \text{otherwise} \end{cases} \tag{5.21}$$

$$W_4(v) = \frac{T}{2} \cdot \frac{1.08 - 0.16T^2v^2}{1 - T^2v^2}\mathrm{sinc}(Tv). \tag{5.22}$$

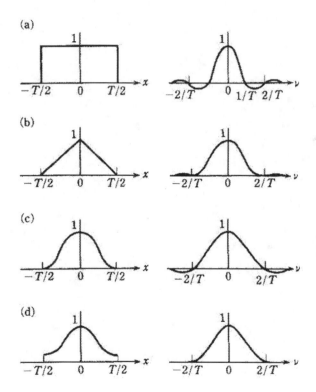

Figure 5.2 Window functions and their spectrum. (a) Rectangular window, (b) Bartlett window, (c) Hanning window and (d) Hamming window.

5.3 PRINCIPLE OF FAST FOURIER TRANSFORM (FFT)

Consider again the discrete Fourier transform of Eq. (5.10),

$$F(k) = \sum_{n=0}^{N-1} f(n)W^{kn}, \tag{5.23}$$

Table 5.1

Decomposition of Indices of k and n ($r_1 = 4$, $r_2 = 3$)

k/n	k_1	k_0	n_1	n_2
0	0	0	0	0
1	0	1	0	1
2	0	2	0	2
3	0	3	1	0
4	1	0	1	1
5	1	1	1	2
6	1	2	2	0
7	1	3	2	1
8	2	0	2	2
9	2	1	3	0
10	2	2	3	1
11	2	3	3	2

where $k = 0, 1, 2, \cdots, N - 1$ and

$$W = \exp\left(-\frac{i2\pi}{N}\right). \tag{5.24}$$

When the discrete Fourier transform is calculated based on Eq. (5.23), multiplication of N^2 times and addition of N^2 times should be performed. Since the calculation time of multiplication is much larger than addition in general, the total calculation time mainly depends on the calculation time of multiplication. The computing time is measured by the multiplication time.

Consider the sampling number of N is factorized by two factors, r_1 and r_2,

$$N = r_1 \times r_2. \tag{5.25}$$

The index of the spectrum k is described by

$$k = k_1 r_1 + k_0, \tag{5.26}$$

where

$$k_0 = 0, 1, 2, \cdots, r_1 - 1,$$

$$k_1 = 0, 1, 2, \cdots, r_2 - 1.$$

As shown in Table 5.1, k is decomposed into r_2 groups of k_1 members.

In the same manner, the index n of input data is described by two factors, n_1 and n_0,

$$n = n_1 r_2 + n_0, \tag{5.27}$$

where

$$n_0 = 0, 1, 2, \cdots, r_2 - 1$$
$$n_1 = 0, 1, 2, \cdots, r_1 - 1.$$

By using these decomposition, Eq. (5.10) is written by

$$F(k_1, k_0) = \sum_{n_1=0}^{r_1-1} \sum_{n_0=0}^{r_2-1} f(n_1, n_0) W^{kn_1 r_2} W^{kn_0}. \tag{5.28}$$

Then the summation for n_1 is rewritten by

$$F_1(k_0, n_0) = \sum_{n_1=0}^{r_1-1} f(n_1, n_0) W^{kn_1 r_2}. \tag{5.29}$$

Equation (5.29) is Fourier transform of a sequence given by decimating N data array $f(n_1, n_0)$ every r_2. Furthermore, submitting Eq. (5.29) to Eq. (5.28) gives

$$F(k_1, k_0) = \sum_{n_0=0}^{r_2-1} F_1(k_0, n_0) W^{(k_1 r_1 + k_0) n_0}. \tag{5.30}$$

This is also DFT. This means that a DFT can be decomposed into two steps of DFTs. In Eq. (5.29), $F_1(k_0, n_0)$ consists of N components. Since each component needs r_1 time multiplications, Nr_1 multiplications are necessary to calculate $F_1(k_0, n_0)$ in total.[2] The calculation of two step DFT needs $N(r_1 + r_2)$.

The two step decomposition reduces the computation time by $N(r_1 + r_2)/N^2 = (r_1 + r_2)/N$ times. It should be noted that the further reduction in computational time can be performed by the decomposition of r_1 and r_2. Based on this idea, the computational time can be reduced drastically. This is the algorithm of the fast Fourier transform (FFT) [1].

As the sample number of N in FFT, 2^m, the power of 2, is selected in many cases. In the case when the calculation is decomposed into m steps, the computational time is estimated as

$$C = N(2 + 2 + \cdots + 2) = N \times 2m = 2N \log_2 N. \tag{5.31}$$

A comparison of the computation time between the cases of direct DFT and FFT is shown in Fig. 5.3. More the increase of the sample number, more the power of FFT is displayed.

---

[2]To calculate the index of W, multiplication of $k_0 n_1 r_2$ is necessary, but this is integer multiplication, for which calculation time is short. Here the complex multiplication is considered.

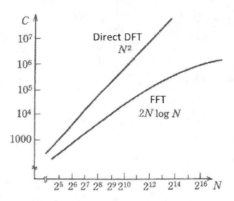

Figure 5.3   Number of operations required computing discrete Fourier transform using FFT
algorithm compared with its direct one.

5.4   NUMERICAL CALCULATION USING FFT

Many software packages of FFT have been developed in many programming lan-
guages, including FORTRAN, C, Basic, VBA, Mathematica, Maple, Python, and
so on. An example of Python code for Fourier transforming of a linear sequence is
presented here. Python is easy-to-use, powerful and open access object-oriented programming language. As well as a rich library, there are several thousand components
and modules available.

A very simple example of the Fourier transform by Python is shown as follows:

```
1     import numpy as np
2     import matplotlib.pyplot as plt
3
4     # rect function
5     def rect(x):
6         return(np.where(np.abs(x)<=0.5,1,0))
7
8     N = 32 # sample number
9     x = np.linspace(-N/2-1,N/2,N)
10    # input signal: rectangular with spacing of 5
11    sig = rect(x/5)
12
13    fig = plt.figure()
14    plt.plot(x,sig) # plot of input signal
15    plt.savefig('fig_rect.jpg') # save plot
16
17    fx = np.fft.fft(sig) # FFT of signal
18    fig = plt.figure()
19    plt.plot(np.real(fx)) # plot of real part of spectrum
20    plt.plot(np.imag(fx),'- -') # imaginary part of spectrum
21    plt.savefig('fig_rect_fft.jpg')
22
```

```
23      sig_shift=np.fft.fftshift(sig) # shift of imput signal
24      fig = plt.figure()
25      plt.plot(x,sig_shift) # plot of shifted signal
26      plt.savefig('fig_rect_shift.jpg')
27
28      fx_shift=np.fft.fft(sig_shift) # FFT of shifted signal
29      fig=plt.figure()
30      plt.plot(np.real(fx_shift)) # plot of real part of spectrum
31      plt.plot(np.imag(fx_shift), '- -') # imaginary part of signal
32      plt.savefig('fig_rect_shift_fft.jpg')
33
34      abs_fx_shift=np.abs(fx_shift)
35      fig=plt.figure()
36      plt.plot(abs_fx_shift**2) # plot of powerspectrum
37      plt.savefig('fig_rect_shift_fft_Intn.jpg')
38
39      fig=plt.figure()
40      # plot of power spectrun shifted to center
39      plt.plot(x,np.fft.fftshift(abs_fx_shift**2))
39      plt.savefig('fig_fft_Intn_shft.jpg)
```

A Python module, Numpy, is used for array computing. Consider a rect function locates in the center of data, as shown in Fig. 5.4(a). In documents in optics, including this book, the object is defined from $-N/2+1$ to $N/2$, while it should be noted that DFT with N data points is defined from $n = 0$ to $N - 1$, according to Eq. (5.10), as shown in Fig. 5.4(b). The real and imaginary parts of Fourier transforms of (a) and (b) are shown in Figs. 5.4(c) and (d), respectively. The solid lines and dotted lines correspond to their real and imaginary parts. The spectrum intensity of Figs. 5.4 (c) and (d) is shown in Fig. 5.4 (e) and its shifted version in Fig. 5.4(f).

The 2-D DFT is defined as

$$F(k, j) = \sum_{m=0}^{N-1} \sum_{n=0}^{N-1} f(m,n) W^{km} W^{jn}. \tag{5.32}$$

The 1-D DFT to the x-direction is given by

$$F_x(k,n) = \sum_{m=0}^{N-1} f(m,n) W^{km}. \tag{5.33}$$

Then, Eq. (5.32) can be separated to

$$F(k, j) = \sum_{n=0}^{N-1} F_x(k,n) W^{jn}. \tag{5.34}$$

This means that the 2-D DFT can be made by sequential 1-D DFTs to the x-direction and then to the y-direction.

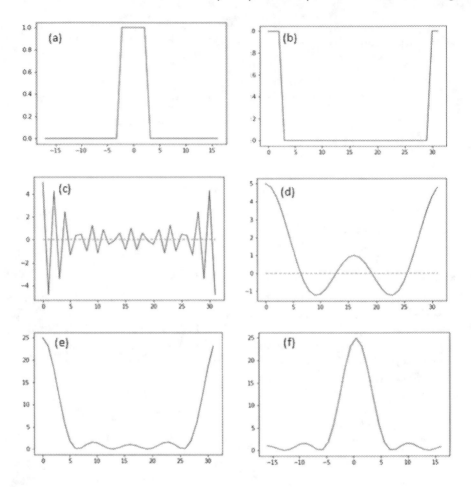

Figure 5.4 Numerical calculation using FFT. $N = 32$. (a) Input data, (b) shifted input data, (c) FFT of (a), real part (solid line) and imaginary part (dotted line), (d) FFT of (b), real part (solid line) and imaginary part (dotted line), (e) intensity of (c) and (d), and (f) shifted intensity of (e).

5.5   INTERPOLATION IN DFT

5.5.1   ZERO PADDING

One of the common techniques for increasing the sampling rate in the spectrum domain is "zero padding." The zero padding is to append zero-valued samples to the end of the original samples. Consider the DFT of a sequence of N samples, $f(p\Delta)$,

$0 \leq p \leq N - 1$, as follows,

$$F(n\Omega) = \sum_{p=0}^{N-1} f(p\Delta) \exp(-inp\Omega\Delta), \qquad (5.35)$$

where the sampling period is denoted by Δ and $\Omega = 2\pi/(N\Delta)$. With this specification of n, there are only N distinct and independent values computable by Eq. (5.35), namely, those for $n = 0, ..., N - 1$.

Set K subperiods inside the sampling period Ω. The new period is given by $\omega = \Omega/K$. The sequence $F(n\Omega)$ provides samples of the desired sequence only at intervals of K samples at the new sampling rate. The remaining samples must be filled in by interpolation [2]. The shifting property in the discrete Fourier transform is

$$F(n\Omega + k\omega) = \sum_{p=0}^{N-1} f(p\Delta) \exp(-ikp\omega\Delta) \exp(-inp\Omega\Delta), \qquad (5.36)$$

$$(k = 0, 1, ..., K - 1).$$

Equation (5.37) is described by the N-point DFT of the sequence $f(p\Delta)$ with a phase factor $\exp(-ikp\omega\Delta)$. The interpolated sample $F(n\Omega + k\omega)$ is given by using an appropriate phase factor $\exp(-ikp\omega\Delta)$. The total sequence of NK samples is obtained by successive interpolation procedures.

The inverse DFT of the interpolated sequence $F(n\Omega + k\omega)$ is considered here, which is given by

$$f'(r\Delta) = \frac{1}{NK} \sum_{n=0}^{N-1} \sum_{k=0}^{K-1} F(n\Omega + k\omega)] \exp[i(n\Omega + k\omega)r\Delta], \qquad (5.37)$$

$$(r = 0, 1, ..., NK - 1).$$

Substituting Eq. (5.37) into Eq. (5.38) we have

$$f'(r\Delta) = \frac{1}{NK} \sum_{n=0}^{N-1} \sum_{k=0}^{K-1} \sum_{p=0}^{N-1} f(p\Delta) \exp(-ikp\omega\Delta)$$

$$\exp(-inp\Omega\Delta) \exp[i(n\Omega + k\omega)r\Delta]$$

$$= \begin{cases} f(r\Delta): & 0 \leq r \leq N - 1 \\ 0: & N \leq r \leq NK - 1. \end{cases} \qquad (5.38)$$

From Eq. (5.38), DFT of the interpolated spectrum $F(n\Omega + k\omega)$ is the original N-sampled sequence $f(r\Delta)$ with zero-valued samples. This procedure suggests that the Fourier transform of a sample sequence of the original sample sequence appended with a zero-valued sample gives the interpolated Fourier spectrum of the original Fourier one [2]. This procedure is called zero-padding.

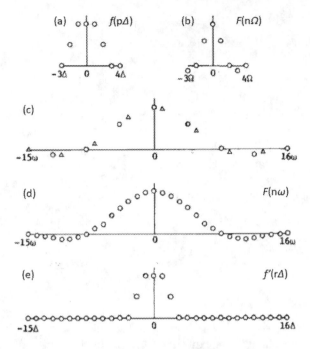

Figure 5.5 Interpolation method in one-dimensional (1-D) case (N = 8, K = 4). (a) Original sequence $f(p\Delta), 0 \leq p < 8$, (b) its DFT $F(n\Omega)$, (c) sequence in interpolation (○: original sequence, △: first interpolated sequence), (d) interpolated sequence $F(n\Omega + k\omega)$, and (e) sequence $f'(r\Delta), 0 \leq r \leq 32$, which is given by the inverse DFT of (d).

5.5.2 SOME OTHER INTERPOLATION TECHNIQUES

In Sec. 5.5.1, it is discussed that zero-padding enables interpolation in the object domain. In this section, some other interpolation techniques are discussed based on DFT [3].

Assume the sampling period Δ in the object domain is divided into K-subperiod is given by $\delta = \Delta/K$. Since the sequence $f(\Delta p)$ provides samples of the desired function only at intervals of K samples at a new sampling rate, we must fill in the remaining samples by interpolation. Next, we define the k-th subsequence made up of the k-th sample in each period of Δ. The interpolated sequence $f(p\Delta + k\delta)$ is defined according to some interpolation type. The inverse DFT of the interpolated sequence $F(p\Delta + k\delta)$ is given by

$$F'(r\Omega) = \frac{1}{NK} \sum_{p=0}^{N-1} \sum_{k=0}^{K-1} f(p\Delta + k\delta) \exp[-i(p\Delta + k\delta)r\Omega], (r = 0, 1, \cdots, NK - 1).$$

(5.39)

Interpolated Fourier spectrum is given by substituting $f(p\Delta + k\delta)$ into Eq. (5.39).

There are various types of interpolation, such as constant, triangle, bell and cubic B-spline interpolation[3]. Here, we consider only the constant interpolation or the zero-order interpolation. The number of spectrum points increases because one sample of the interpolated spectrum is made by K-time interlace. The interpolated sequence is $f(p\Delta+k\delta) = f(p\Delta)$, then Eq. (5.39) is rewritten as

$$
\begin{aligned}
F'(r\Omega) &= \frac{1}{NK} \sum_{p=0}^{N-1} \sum_{k=0}^{K-1} f(p\Delta)\exp[-i(p\Delta+k\delta)r\Omega] \\
&= \frac{1}{NK} \sum_{p=0}^{N-1} \sum_{k=0}^{K-1} \sum_{n=0}^{N-1} F(n\Omega)\exp(inp\Omega\Delta)\exp[-i(p\Delta+k\delta)r\Omega] \\
&= F([r]_N\Omega)W(r),
\end{aligned}
\tag{5.40}
$$

where $[r]_N = r$ modulo N, and $W(r)$ is the weight factor in the Fourier domain given by

$$
W(r) = \frac{1}{K} \frac{1-\exp(-i2\pi r/N)}{1-\exp(-i2\pi r/NK)}.
\tag{5.41}
$$

The result is the product of the sequence $F([r]_N\Omega)$ which is the periodic extension of $F(n\Omega)$ and the weight factor $W(r)$.

PROBLEMS

1. Derive Eq. (5.5).
2. Prove the linearity, similarity and shift theorem in DFT.
3. Prove the convolution theorem, correlation theorem and the Parseval formula in DFT.
4. Compute $\text{rect}(x)$ and $\text{rect}(x-a)$ using FFT, and graph their real and imaginary parts.
5. Compute the convolution of $\text{rect}(x)$ and $\text{rect}(x/a)$ using FFT.
6. Compare the computational times for the autocorrelation with sample number of $N = 2^M$.
7. The computation time for a real data sequence with the sample number N can be reduced by dividing the data into the odd number and even number data sequence.
   For the real data sequence $f(n), (n = 0, 1, ..., N-1)$, divide into two-sequences

   $$
   g_1(m) = f(2m), \quad g_2(m) = f(2m+1) \quad (m = 0, 1, ..., N/2-1).
   $$

   Then, a complex sequence of the data number $N/2$

   $$
   h(m) = g_1(m) + ig_2(m)
   $$

   is made. Show the real sequence $f(n)$ can be Fourier transformed by FFT of the complex sequence $h(m)$ with the data number $N/2$.

BIBLIOGRAPHY

Nussbaumer, H. J. 1981. *Fast Fourier Transform and Convolution Algorithms*. Springer-Verlagl, Berlin.

Landau, R. H., Paez, M. J. and Bordeianu, C. C. 2015. *Computational Physics: Problem Solving with Python*, 3rd ed. Wiley-VCH Verlag GmbH & Co.

REFERENCES

1. Cooley, J. W. and Tukey, J. W. 1965. An algorithm for the machine calculation of complex Fourier series. *Math. Comput.* 19: 297.
2. Yatagai, T. 1977. Interpolation method of computer-generated filters for large object formats. *Opt. Commun.* 15: 347.
3. Yoshikawa, N., Itoh M. and Yatagai, T. 1995. *Interpolation of reconstructed image in Fourier transform computer-generated hologram. Opt. Commun.* 119: 33.

6 Fourier Optics

Optical systems can be considered as linear systems, which, therefore, can be analyzed by using Fourier transforms. At first, we show that the Fresnel diffraction is described as a convolution integral, and then the Fourier transform property of lens is discussed. Systematic understanding of many optical characteristics of coherent and incoherent imaging systems is performed. The optical transfer functions (OTF) in the coherent and incoherent imaging systems are introduced to discuss the imaging properties. Then the resolving power is defined. The angular spectrum method for wave propagation analysis is also discussed. Finally, the diffraction in 3-D space based on 3-D spectrum is presented.

6.1 FRESNEL DIFFRACTION

Consider again, the Fresnel diffraction equation of (2.41). In the optical setup of Fig. 6.1, let the amplitudes in the planes P_1 and P_2 be $f(x_i, y_i)$ and $g(x_0, y_0)$, respectively. If the distance l satisfies the Fresnel diffraction condition, Eq. (2.39), the amplitude $g(x_0, y_0)$ is given by

$$g(x_0, y_0) = \frac{A}{i\lambda l} \exp(ikl) \iint_{-\infty}^{\infty} f(x_i, y_i) \exp\left\{\frac{ik}{2l}[(x_i - x_0)^2 + (y_i - y_0)^2]\right\} dx_i dy_i. \quad (6.1)$$

Equation (6.1) is simply rewritten by the convolution integral

$$g(x_0, y_0) = Af * h_l(x_0, y_0), \quad (6.2)$$

where

$$h_l(x_0, y_0) = \frac{1}{i\lambda l} \exp\left[i\frac{\pi}{\lambda l}(x_0^2 + y_0^2)\right], \quad (6.3)$$

which is called the impulse response or the point spread function in optics.

Equation (6.2) shows that a linear system with the response function $h_l(x_0, y_0)$ gives the output $g(x_0, y_0)$ for the input $f(x_i, y_i)$. The Fourier transform of Eq. (6.2) is given by

$$G(\nu_x, \nu_y) = C \cdot F(\nu_x, \nu_y) \cdot H(\nu_x, \nu_y), \quad (6.4)$$

where

$$G(\nu_x, \nu_y) = \mathscr{F}[g(x_0, y_0)] \quad (6.5)$$

$$F(\nu_x, \nu_y) = \mathscr{F}[f(x_i, y_i)] \quad (6.6)$$

$$H(\nu_x, \nu_y) = \mathscr{F}[h_l(x_0, y_0)] = \exp[-i\lambda l\pi(\nu_x^2 + \nu_y^2)] \quad (6.7)$$

and C denotes a constant. The system diagram of the Fresnel transform is shown in Fig. 6.2.

DOI: 10.1201/9781003121916-6

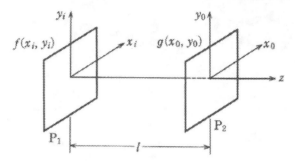

Figure 6.1 Geometry for Fresnel diffraction calculation.

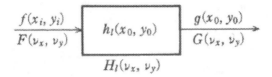

Figure 6.2 Optical system of Fresnel diffraction.

According to the impulse response, it should be noted that the system output $h_l(x_0, y_0)$ is given by a point source input $\delta(x_i, y_i)$ located at Plane P_1, and Eq. (6.3) is a spherical wave, described by Eq. (1.35).[1]

6.2　FOURIER TRANSFORM OPERATION OF LENS

Consider a point source Q located in front of a lens at a distance of a and its image P located at a distance b from the lens, as shown in Fig. 6.3. The lens imaging equation is given by

$$\frac{1}{a} + \frac{1}{b} = \frac{1}{f}, \tag{6.8}$$

where f denotes a focal length. The wavefront from a point source Q, arriving at the front plane of the lens is a spherical wave given by

$$u^-(x, y) = A \exp\left[i\frac{\pi}{\lambda a}(x^2 + y^2)\right]. \tag{6.9}$$

In the same manner, a spherical wavefront from the backplane of the lens with an aperture $p(x, y)$, arriving at a point image is given by

$$u^+(x, y) = A' \exp\left[-i\frac{\pi}{\lambda b}(x^2 + y^2)\right] \cdot p(x, y), \tag{6.10}$$

---

[1] It is not correctly spherical but a paraboloid of revolution.

where the function $p(x,y)$ is called the pupil function defined as

$$p(x,y) = \begin{cases} 1 : \text{inside of the lens aperture} \\ 0 : \text{outside of the lens aperture.} \end{cases} \tag{6.11}$$

The transmittance $t(x,y)$ of the lens is defined as

$$t(x,y) = \frac{u^+(x,y)}{u^-(x,y)}, \tag{6.12}$$

where the amplitude is not changed between the front plane A and backplane A' of the lens. By using Eq. (6.8), we have

$$t(x,y) = \exp\left[-i\frac{\pi}{\lambda f}(x^2+y^2)\right] \cdot p(x,y). \tag{6.13}$$

This equation describes the lens operation, whose operational diagram is shown in Fig. 6.4.

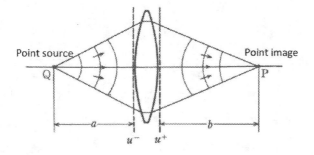

Figure 6.3 Operation of lens.

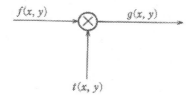

Figure 6.4 Linear system of lens.

Consider an object $f(x,y)$ located in front of the lens at a distance l and the observation plane is at the back focal plane of the lens, as shown in Fig. 6.5. Since the wave propagation of a distance l from the object plane is Fresnel diffraction, the

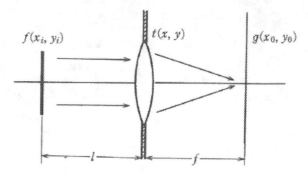

Figure 6.5   Fourier transform operation of lens.

wavefront just in front of the lens is given as

$$u^-(x,y) = f(x,y) * \frac{1}{i\lambda l} \exp\left[i\frac{\pi}{\lambda l}(x^2 + y^2)\right] \tag{6.14}$$

$$= \frac{1}{i\lambda l} \iint f(x_i, y_i) \exp\left\{i\frac{\pi}{\lambda l}\left[(x - x_i)^2 + (y - y_i)^2\right]\right\} dx_i dy_i$$

by using Eq. (6.2). Its Fourier transform is

$$U^-(v_x, v_y) = F(v_x, v_y) \exp[-i\lambda l\pi(v_x^2 + v_y^2)]. \tag{6.15}$$

The amplitude just behind the lens is given by

$$u^+(x,y) = t(x,y)u^-(x,y). \tag{6.16}$$

Fresnel transform of Eq. (6.16) gives the amplitude in the focal plane

$$g(x_0, y_0) = u^+(x,y) * \frac{1}{i\lambda f} \exp\left[i\frac{\pi}{\lambda f}(x^2 + y^2)\right] \tag{6.17}$$

$$= \frac{1}{i\lambda f} \iint u^+(x,y) \exp\left\{i\frac{\pi}{\lambda f}\left[(x_0 - x)^2 + (y_0 - y)^2\right]\right\} dx dy.$$

Finally, by inserting Eqs. (6.16) and (6.13), we have

$$g(x_0, y_0) = \frac{1}{i\lambda f} \exp\left[i\frac{\pi}{\lambda f}(x_0^2 + y_0^2)\right]$$

$$\times \iint u^-(x,y)p(x,y) \exp\left[-i\frac{2\pi}{\lambda f}(xx_0 + yy_0)\right] dx dy$$

$$= \frac{1}{i\lambda f} \exp\left[i\frac{\pi}{\lambda f}(x_0^2 + y_0^2)\right] U^-(v_x, v_y) * P(v_x, v_y), \tag{6.18}$$

where

$$v_x = \frac{x_0}{\lambda f} \tag{6.19}$$

$$v_y = \frac{y_0}{\lambda f} \tag{6.20}$$

$$U^-(v_x, v_y) = \mathscr{F}[u^-(x,y)] \tag{6.21}$$

$$P(v_x, v_y) = \mathscr{F}[p(x,y)]. \tag{6.22}$$

Consider the case when the aperture size is large enough, that is, $p(x,y) = 1$.

$$P(v_x, v_y) = \delta(v_x, v_y). \tag{6.23}$$

Finally, we have

$$g(x_0, y_0) = \frac{1}{i\lambda f} \exp\left[i\frac{\pi}{\lambda f}\left(1 - \frac{l}{f}\right)(x_0^2 + y_0^2)\right] \cdot F\left(\frac{x_0}{\lambda f}, \frac{y_0}{\lambda f}\right). \tag{6.24}$$

This means Fourier transform of the input signal $f(x,y)$ is obtained in the focal plane, except for the phase term $\exp[i\pi(1 - l/f)(x_0^2 + y_0^2)/\lambda f]$. It should be noted that the phase term is eliminated in the case of $l = f$. That is, the input located in the front focal plane gives the perfect Fourier transform, which is obtained in the back focal plane. This is called Fourier transform operation of a lens, where the Fourier transform is scaled by $1/\lambda f$.

6.3 COHERENT IMAGING

Until now, we have implicitly discussed diffraction phenomena for a monochromatic light source and light waves from object points are coherent with each other. In the same conditions, imaging phenomena are analyzed in this chapter. Assume that wave propagation is linear and, therefore. The principle of superposition of wave amplitude is valid. In the case of imaging, this principle is valid.

Consider the optical system shown in Fig. 6.6. An object located in the plane P_1 and its image is formed in the plane P_2. When the correspondence of a point (x_i, y_i) on the object to the observation plane is $h(x_0, y_0; x_i, y_i)$ and the amplitude of the object is $f(x_i, y_i)$, the amplitude in the observation plane P_2 is written by the superposition integral as

$$g(x_0, y_0) = \iint f(x_i, y_i) h(x_0, y_0; x_i, y_i) dx_i dy_i. \tag{6.25}$$

The function $h(x_0, y_0; x_i, y_i)$ is the impulse response and called the point response function or the point spread function. If the optical system is perfect, the point image distribution is a perfect point, which is described by the delta function

$$h(x_0, y_0; x_i, y_i) = g(x_0, y_0) = A\delta(x_0 \pm mx_i, y_0 \pm my_i), \tag{6.26}$$

where A and m denote a complex constant and the magnification of the optical system, respectively. In the cases of $+$, an erect image is obtained, and $-$ an inverted image.

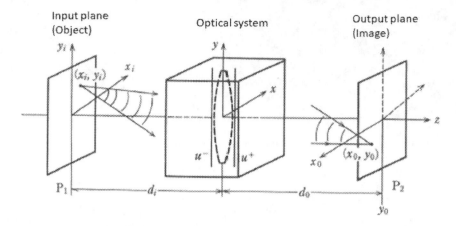

Figure 6.6 Imaging optical system.

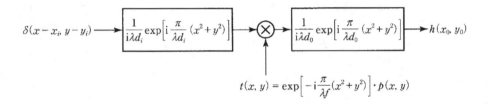

Figure 6.7 Linear system of point object imaging.

In general cases, the point image is not a point because of the diffraction effect of a limited aperture of the imaging optical system.

Next, we calculate the point spread function in a coherent imaging system. Consider the case when a point object located at (x_i, y_i) in the object plane P_1 at a distance of d_i from the imaging lens and is illuminated by monochromatic light with its amplitude of unity, as shown in Fig. 6.6. The complex amplitude at the plane P_1 is described by $\delta(x - x_i, y - y_i)$. The point spread function $h(x_0, y_0; x_i, y_i)$ is given by the amplitude in the image plane P_2 located at a distance of d_0 from the imaging lens. The wavefront $f(x_i, y_i)$ propagates to the lens, then passes through the lens and propagates to the imaging plane. The system diagram of this process is shown in Fig. 6.7, and

$$u^-(x,y) = \delta(x - x_i, y - y_i) * \frac{1}{i\lambda d_i} \exp\left[i\frac{\pi}{\lambda d_i}(x^2 + y^2)\right] \tag{6.27}$$

$$u^+(x,y) = t(x,y) \times u^-(x,y) \tag{6.28}$$

$$g(x_0, y_0) = u^+(x_0, y_0) * \frac{1}{i\lambda d_0} \exp\left[i\frac{\pi}{\lambda d_0}(x_0^2 + y_0^2)\right]. \tag{6.29}$$

Finally we have

$$
\begin{aligned}
h(x_0,y_0;x_i,y_i) =& g(x_0,y_0) \\
=& \frac{1}{\lambda^2 d_i d_0} \exp\left[i\frac{\pi}{\lambda d_0}(x_0^2+y_0^2)\right] \exp\left[i\frac{\pi}{\lambda d_i}(x_i^2+y_i^2)\right] \\
& \times \iint p(x,y) \exp\left[i\frac{\pi}{\lambda}\left(\frac{1}{d_i}+\frac{1}{d_0}-\frac{1}{f}\right)(x^2+y^2)\right] \\
& \times \exp\left\{-i\frac{2\pi}{\lambda}\left[\left(\frac{x_i}{d_i}+\frac{x_0}{d_0}\right)x+\left(\frac{y_i}{d_i}+\frac{y_0}{d_0}\right)y\right]\right\}dxdy.
\end{aligned}
\tag{6.30}
$$

Since the intensity of the image is only observable, the phase term outside the integral $\exp[i\pi/\lambda d_0 \cdot (x_0^2+y_0^2)]$ may be neglected, but the other phase term $\exp[i\pi/\lambda d_i \cdot (x_i^2+y_i^2)]$ can not be neglected, because the integral variables x_i and y_i are included in the superposition integral Eq. (6.26).

In a normal imaging system, all the area of the object plane does not contribute the one point in the imaging plane, but the limited area of the object forms the image point. Accordingly, we can consider $x_i = x_0/m$ and $y_i = y_0/m$ and, therefore this phase term is also considered as a function of (x_0,y_0). Due to the same reason, when the first phase term is neglected, the second phase term can be neglected. Equation (6.30) is rewritten as

$$
\begin{aligned}
h(x_0,y_0;x_i,y_i) =& \frac{1}{\lambda^2 d_i d_0} \iint p(x,y) \exp\left[i\frac{\pi}{\lambda}\left(\frac{1}{d_i}+\frac{1}{d_0}-\frac{1}{f}\right)(x^2+y^2)\right] \\
& \times \exp\left\{-i\frac{2\pi}{\lambda}\left[\left(\frac{x_i}{d_i}+\frac{x_0}{d_0}\right)x+\left(\frac{y_i}{d_i}+\frac{y_0}{d_0}\right)y\right]\right\}dxdy.
\end{aligned}
\tag{6.31}
$$

By using the lens equation

$$
\frac{1}{d_i}+\frac{1}{d_0}-\frac{1}{f}=0
\tag{6.32}
$$

we have

$$
h(x_0,y_0;x_i,y_i) = \frac{1}{\lambda^2 d_i d_0} \iint p(x,y) \exp\left\{-i\frac{2\pi}{\lambda}\left[\left(\frac{x_i}{d_i}+\frac{x_0}{d_0}\right)x+\left(\frac{y_i}{d_i}+\frac{y_0}{d_0}\right)y\right]\right\}dxdy.
\tag{6.33}
$$

Since the magnification m of the imaging system is given by

$$
m = \frac{d_0}{d_i}.
\tag{6.34}
$$

Equation (6.33) is rewritten as

$$
h(x_0,y_0;x_i,y_i) = \frac{1}{\lambda^2 d_i d_0} \iint p(x,y) \exp\left\{-i\frac{2\pi}{\lambda d_0}[(x_0+mx_i)x+(y_0+my_i)y]\right\}dxdy.
\tag{6.35}
$$

By defining the new coordinates

$$v_x = \frac{x}{\lambda d_0} \tag{6.36}$$

$$v_y = \frac{y}{\lambda d_0}. \tag{6.37}$$

The point spread function is given by

$$h(x_0, y_0; x_i, y_i) = \iint p(\lambda d_0 v_x, \lambda d_0 v_y) \exp\{-i2\pi[(x_0 + m x_i)v_x + (y_0 + m y_i)v_y]\} dv_x dv_y. \tag{6.38}$$

This means that the point spread function is a Fourier transform of the pupil function $p(\lambda d_0 v_x, \lambda d_0 v_y)$. By further coordinate transform

$$x_i' = -m x_i \tag{6.39}$$

$$y_i' = -m y_i. \tag{6.40}$$

The point spread function is a function of $x_0 - x_i'$ and $y_0 - y_i'$, given by

$$h(x_0 - x_i', y_0 - y_0') = \iint p(\lambda d_0 v_x, \lambda d_0 v_y) \exp\{-i2\pi[(x_0 - x_i')v_x + (y_0 - y_i')v_y]\} dv_x dv_y. \tag{6.41}$$

This point spread function is a function of only the differences of coordinates and therefore its shape is invariant, not depending on the object position but on the relative position. This means that this optical system is space-invariant and the superposition integral of Eq. (6.25) is written as a convolution integral,

$$g(x_0, y_0) = \frac{1}{m^2} \iint f\left(-\frac{x_i'}{m}, -\frac{y_i'}{m}\right) h(x_0 - x_i', y_0 - y_i') dx_i' dy_i'. \tag{6.42}$$

Here we redefine coordinates in the plane P_2 as (x_i, y_i) according to Eqs. (6.39) and (6.40) and redefine $h(x_0 - x_i', y_0 - y_i')/m$ and $f(-x_i/m, -y_i/m)/m$ as $h(x_0 - x_i, y_0 - y_i)$ and $f(x_i, y_i)$, respectively. We have

$$g(x_0, y_0) = \iint f(x_i, y_i) h(x_0 - x_i, y_0 - y_i) dx_i dy_i = f * h(x_0, y_0). \tag{6.43}$$

The intensity of the image is given by

$$I_c = |f * h(x_0, y_0)|^2. \tag{6.44}$$

In the case when the wavelength λ is sufficiently small and the size of the pupil is sufficiently large, we can assume $p = 1$. From Eq. (6.41) we have

$$h(x_0, y_0; x_i, y_i) = \delta(x_0 - x_i, y_0 - y_i). \tag{6.45}$$

Only in this case, the effect of the diffraction is negligible and the perfect image is obtained.

6.4 INCOHERENT IMAGING

When an object is illuminated by a white light source not by a special source, like laser, the light waves scattered from the object are incoherent with each other. Let the wave amplitude $f(x_i, y_i)$ and $f(x_i', y_i')$ from the points (x_i, y_i) and (x_i', y_i') on the object. Referring Sec. 10.1, if the two waves are incoherent, the time average of the product of the two waves $\langle f(x_i, y_i) f^*(x_i', y_i') \rangle$ can be written as

$$\langle f(x_i, y_i) f^*(x_i', y_i') \rangle = c I_f(x_i, y_i) \delta(x_i - x_i', y_i - y_i'), \tag{6.46}$$

where c denotes a constant. This means that the intensity $I_f(x_i, y_i)$ is obtained due to the interference only between the waves from the same point on the object and the intensity in other cases is zero because waves from the different points do not interfere with each other.

In this case, the intensity of the image is given by

$$I_i(x_0, y_0) = c \iint |h(x_0 - x_0', y_0 - y_0')|^2 I_f(x_0', y_0') \mathrm{d}x_0' \mathrm{d}y_0'. \tag{6.47}$$

The intensity of the image I_i is written by a convolution integral of the intensity of the point spread function $|h|^2$ and the intensity of the object I_f. This relation corresponds to the relation that the amplitude of the image g is given by the convolution integral of the amplitude of the point spread function h and the amplitude of the object f in the coherent imaging. The intensity of the image in the coherent imaging is given by the squared absolute value of the amplitude of the image. The imaging system in each case is summarized as follows;

Coherent imaging: $I_c = |h * f|^2$

Incoherent imaging: $I_i = |h|^2 * I_f = |h|^2 * |f|^2$.

6.5 FREQUENCY RESPONSE OF OPTICAL SYSTEM

Consider optical transfer functions (OTF) in coherent and incoherent imaging. The frequency response includes all the information about the linear system, as mentioned Secs. 4.2.3 and 4.3.

6.5.1 COHERENT IMAGING

Since the convolution of amplitudes of the object and the point spread function gives the amplitude of the image, it is reasonable that the frequency response function is defined by Fourier transform of the amplitude of the point response function. From Eq. (6.41) and with $m = 1$, we have

$$h(x_0, y_0) = \iint p(\lambda d_0 v_x, \lambda d_0 v_y) \exp[-i2\pi(x_0 v_x + y_0 v_y)] \mathrm{d}v_x \mathrm{d}v_y. \tag{6.48}$$

Fourier transforming the point spread function gives the frequency response

$$\begin{aligned} H(v_x, v_y) &= \mathscr{F}[h(x_0, y_0)] \\ &= \mathscr{F}\{\mathscr{F}[p(\lambda d_0 v_x, \lambda d_0 v_y)]\} \\ &= p(-\lambda d_0 v_x, -\lambda d_0 v_y). \end{aligned} \tag{6.49}$$

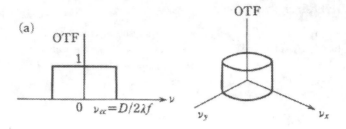

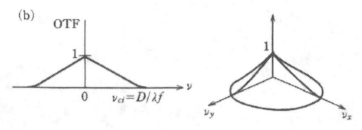

Figure 6.8 Optical transfer functions in perfect coherent imaging (a) and perfect incoherent imaging (b).

By defining coordinates of the pupil function appropriately, we have

$$H(v_x, v_y) = p(\lambda d_0 v_x, \lambda d_0 v_y). \tag{6.50}$$

The OTF in the coherent imaging system is equal to the pupil function itself. Figure 6.8(a) shows OTF of a perfect coherent imaging system with a diameter D of a circular pupil. The highest frequency passing through this coherent imaging system called a cut-off frequency, is

$$v_{cc} = \frac{D}{2\lambda f}. \tag{6.51}$$

6.5.2 INCOHERENT IMAGING

In the incoherent imaging system, the image intensity is given by the convolution of the object intensity and the intensity of the point response function. OTF of the incoherent imaging system is given by

$$\boldsymbol{H}(v_x, v_y) = \mathscr{F}[|h(x_0, y_0)|^2]. \tag{6.52}$$

Usually, the OTF is normalized with the value of zero frequency

$$
\begin{aligned}
H(v_x, v_y) &= \frac{\mathscr{F}\left[|h(x_0, y_0)|^2\right]}{\mathscr{F}\left[|h(x_0, y_0)|^2\right]_{v_x=0, v_y=0}} \\
&= \int \int \frac{H(v_x, v_y) \star H^*(v_x, v_y)}{|H|^2 dv_x dv_y} \\
&= \int \int \frac{p(\lambda dv_x, \lambda dv_y) \star p^*(\lambda dv_x, \lambda dv_y)}{|p|^2 dv_x dv_y}.
\end{aligned}
\tag{6.53}
$$

The autocorrelation of the pupil function gives the OTF in the incoherent imaging system.

Figure 6.8(b) shows the OTF of a perfect incoherent imaging system with a diameter D of a circular pupil. Its cut-off frequency is given by

$$
v_{ci} = \frac{D}{\lambda f},
\tag{6.54}
$$

and is two times larger than that of the coherent imaging system with the same pupil dimension.

6.6   RESOLVING POWER

As mentioned above, the concept of OTF is important to evaluate the optical performance. The relation between the OTF and the optical resolution, which is widely used as one of the optical performance measures, is discussed.

The optical resolution is a measure of how fine object an optical system can resolve. It depends on the detection conditions. Because the object is observed by the eye or by an electro-optical detector, it is difficult to define in general.

One of the useful criteria of the resolution limits is the Rayleigh criterion. Consider two point sources incoherent with each other and the imaging system is ideal without aberration. The Rayleigh criterion is defined as the smallest distance between two images of the point sources that can be resolved. As shown in Fig. 6.9, if point sources are sufficiently separated, point sources are resolved as two separated images. If two point sources move closer, their images set close, overlap closing, overlap and finally are observed as one image.

From Eq. (2.67), the image intensity is described as

$$
I(w) = I_0 \left| \frac{2J_1\left(\frac{kD}{2f}w\right)}{\frac{kD}{2f}w} \right|^2,
\tag{6.55}
$$

The Rayleigh criterion gives the minimum distance between two point images which can be resolved as two images. The minimum distance L is defined as the distance between the position of the maximum intensity of a point image and the first minimum intensity of the other point image, the radius Δw of the first dark annular ring of the Airy disc given by

$$
\Delta w = 1.22 \frac{\lambda f}{D}
\tag{6.56}
$$

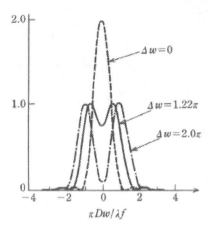

Figure 6.9 Rayleigh criterion. Intensity distribution of two ideal point images with different spacing Δw.

from Eq. (2.69). The minimum distance is given by

$$L = 1.22 \frac{\lambda f}{D}. \tag{6.57}$$

In this case, the intensity of the middle point between two point images is about 20% less than the peak intensity of the point images. The reciprocal of the distance L gives the resolving power R given by

$$R = \frac{1}{L} = 0.82 \frac{D}{\lambda f}. \tag{6.58}$$

The cut-off frequency $v_c = D/\lambda f$ based on the OTF and the resolving power R is well corresponding to each other, but the difference is about 20%. OTF describes the frequency response for all spatial frequencies, but the resolving power R shows only the frequency limit based on the Rayleigh criterion. It should be noted that the cut-off frequency is equal to the resolving power R for the rectangular pupil of the imaging system because the Fraunhofer diffraction of a rectangular aperture is given by Eq. (2.59).

6.7 ANGULAR SPECTRUM METHOD

The Fresnel diffraction equation (2.41) is an approximation equation describing a complex amplitude of the diffraction wavefront observed at a distance from an aperture. The amplitude at a shorter distance from the aperture is given by the Fresnel-Kirchhoff diffraction integral or the Rayleigh-Sommerfeld diffraction equation without approximation.

Another approach is based on Helmholtz equation (1.52). Consider a wave propagating to the z-direction as shown in Fig. 6.10. The aperture locates in the plane

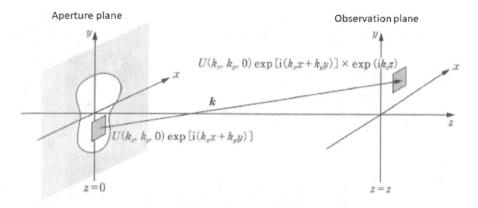

Figure 6.10 Diffraction calculation with angular spectrum method.

$z = 0$ and its coordinates are (x,y). Let us calculate the complex amplitude of the diffracted wave in the plane $z = z$. Fourier transform of the complex amplitude of the wave $u(x,y,0)$ in the aperture plane is given by

$$U(v_x, v_y, 0) = \iint_{-\infty}^{\infty} u(x,y,0) \exp[-i2\pi(xv_x + yv_y)]dxdy \tag{6.59}$$

and its inverse Fourier transform is

$$u(x,y,0) = \iint_{-\infty}^{\infty} U(v_x, v_y, 0) \exp[i2\pi(xv_x + yv_y)]dv_xdv_y. \tag{6.60}$$

This means that the complex amplitude of the wave $u(x,y,0)$ in the aperture plane is decomposed into plane waves $\exp[i2\pi(xv_x, +yv_y)]$ with their amplitude $U(v_x, v_y, 0)$, which propagates to different directions. It should be noted that a plane wave propagating to the k direction is given by $A\exp[i(k \cdot r - \omega t)]$ and its wave number vector components are given by

$$k_x = 2\pi v_x, \qquad k_y = 2\pi v_y, \qquad k_z = \sqrt{k^2 - k_x^2 - k_y^2}. \tag{6.61}$$

Then Eq. (6.59) is rewritten as

$$U(k_x, k_y, 0) = \iint_{-\infty}^{\infty} u(x,y,0) \exp[-i(k_xx + k_yy)]dxdy. \tag{6.62}$$

The term $U(k_x, k_y, z)$ is called the angular spectrum.

To calculate the wave in the observation plane $z = z$, we consider the propagation of the plane wave component $U(k_x, k_y, 0) \exp[i(k_xx + k_yy)]$ from $z = 0$ to $z = z$, which is given by $U(k_x, k_y, 0) \exp[i(k_xx + k_yy)] \times \exp(ik_zz)$.[2]

---

[2]Since the wave at $z = z$ is given by

$$u(x,y,z) = \frac{1}{4\pi^2} \iint_{-\infty}^{\infty} U(k_x, k_y, z) \exp[i(k_xx + k_yy)]dk_xdk_y, \tag{A}$$

Finally, the summation of the propagated component waves gives the complex amplitude of the wave in the observation plane $z = z$

$$u(x,y,z) = \frac{1}{4\pi^2} \iint_{-\infty}^{\infty} U(k_x,k_y,0)\exp[i(k_x x + k_y y)]\exp\left(i\sqrt{k^2 - k_x^2 - k_y^2}\cdot z\right)dk_x dk_y.$$

(6.63)

Figures 6.11 and 6.12 shows examples of diffraction calculation by the angular spectrum method.[3]

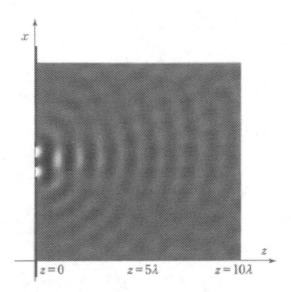

Figure 6.11 Diffraction calculated by angular spectrum method. Double aperture of 0.5λ width and λ spacing in Fig. 6.10.

6.8 DIFFRACTION BASED ON 3-D FOURIER SPECTRUM

Until now, we consider only the 2-D distribution $f(x,y)$ as an input pattern, as is shown in Fig. 6.1. Diffraction from a 3-D distribution $f(x,y,z)$ can be calculated directly based on 3-D Fourier spectrum [3]. Figure 6.13 shows the geometry of

---

this equation should satisfy Helmholtz equation (1.52). By submitting this equation into Eq. (1.52), we have

$$-(k_x^2 + k_y^2)U(k_x,k_y,z) + \frac{d^2}{dz^2}U(k_x,k_y,z) + k^2 U(k_x,k_y,z) = 0.$$

Its solution is

$$U(k_x,k_y,z) = U(k_x,k_y,0)\exp\left(i\sqrt{k^2 - k_x^2 - k_y^2}\cdot z\right).$$ (B)

Equation (6.63) is obtained by substituting Eq. (B) into Eq. (A).

[3]As mentioned in Sec. 1.6, the scalar diffraction theory is not correct near the aperture or very small aperture, the size of which is smaller than several wavelengths. These are only qualitative examples.

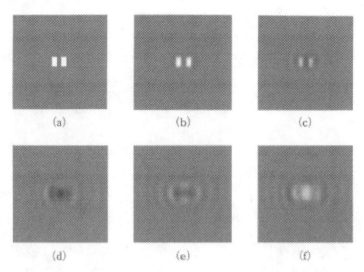

Figure 6.12 Diffraction calculated by angular spectrum method. The same double aperture shown in Fig. 6.11 at (a) $z = 0.0$, (b) $z = 0.1\lambda$, (c) $z = 0.25\lambda$, (d) $z = 0.5\lambda$, (e) $z = 0.75\lambda$, (f) $z = 1.0\lambda$ in Fig. 6.10.

the diffraction from the 3-D object $f(x,y,z)$. Although the 3-D object is illuminated with a coherent light such as a laser, it can be regarded as an aggregation of self-luminous point light sources which emit isotropical spherical wavefronts. The diffraction wavefront at a point (x,y,z) in the 3-D space is given by

$$g(x,y,z) = \iiint f(x_0,y_0,z_0)h(x-x_0,y-y_0,z-z_0)dx_0dy_0dz_0 \qquad (6.64)$$

$$= f * h(x,y,z), \qquad (6.65)$$

where

$$h(x,y,z) = \frac{1}{\sqrt{x^2+y^2+z^2}} \exp\left(\frac{i2\pi}{\lambda}\sqrt{x^2+y^2+z^2}\right). \qquad (6.66)$$

Because the point spread function of Eq. (6.66) represents a spherical wave and it satisfies Helmholtz equation of Eq. (1.52) in 3-D free space, the equation of Eq. (6.64) gives a rigorous diffraction formula. The 3-D Fourier spectrum of $h(x,y,z)$ can be obtained analytically as follows [4]:

$$H(u,v,w) = \mathscr{F}[h(x,y,z)] = \frac{1}{4\pi^2(u^2+v^2+w^2-1/\lambda^2)}, \qquad (6.67)$$

where (u,v,w) denotes the spatial frequency coordinates of (x,y,z). By applying the convolution theorem to Eq. (6.65), Eq. (6.65) is rewritten as

$$g(x,y,z) = \frac{1}{4\pi^2} \iiint \frac{F(u,v,w)}{u^2+v^2+w^2-1/\lambda^2} \exp[i2\pi(ux+vy+wz)]dudvdw, \qquad (6.68)$$

where $F(u,v,w)$ denotes the 3-D Fourier spectrum of $f(x,y,z)$. Because the diffracted wavefront should be considered on a certain 2-D plane, the 2-D distribution of $z = R$ is extracted from $g(x,y,z)$ as follows:

$$g(x,y,z)|_{z=R} = \frac{1}{4\pi^2} \iiint \frac{F(u,v,w)}{u^2 + v^2 + w^2 - 1/\lambda^2} \exp[i2\pi(ux + vy + wR)]dudvdw.$$

(6.69)

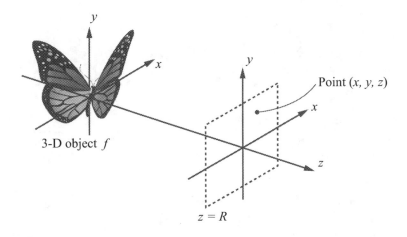

Figure 6.13   Schematic of 3-D viewing optical system.

The integral of Eq. (6.69) with respect to w has two singularities at $w = \pm\sqrt{1/\lambda^2 - u^2 - v^2}$. The integration is made by considering the complex plane $w = \xi + i\eta$ as shown in Fig. 6.14.

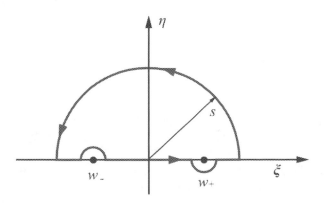

Figure 6.14   Path of line integral.

The path of the line integral on w is a close contour including only one singularity $w_+ = \sqrt{1/\lambda^2 - u^2 - v^2}$. The other singularity $w_- = -\sqrt{1/\lambda^2 - u^2 - v^2}$ is excluded because the wavefront propagating in the $+z$ direction is taken into account here. To calculate the integral with the closed contour, the radius of the hemicycle s must approach infinity. Then, according to the residue theorem, Eq. (6.69) is analytically calculated as

$$
g(x,y,z)|_{z=R} = \frac{i}{4\pi} \iint \frac{F(u,v,\sqrt{1/\lambda^2 - u^2 - v^2})}{\sqrt{1/\lambda^2 - u^2 - v^2}} \exp\left(i2\pi R\sqrt{1/\lambda^2 - u^2 - v^2}\right)
$$
$$
\times \exp[i2\pi(ux+vy)]dudv. \tag{6.70}
$$

Finally, by two-dimensionally Fourier transforming both sides of Eq. (6.70), we have the following equation:

$$
G(u,v)|_{z=R} = \frac{i}{4\pi} \frac{F(u,v,\sqrt{1/\lambda^2 - u^2 - v^2})}{\sqrt{1/\lambda^2 - u^2 - v^2}} \exp\left(i2\pi R\sqrt{1/\lambda^2 - u^2 - v^2}\right). \tag{6.71}
$$

When we define $w(u,v)$ as $\sqrt{1/\lambda^2 - u^2 - v^2}$, the equation $u^2 + v^2 + w(u,v)^2 = 1/\lambda^2$ is satisfied. Therefore, Eq. (6.71) means that the 2-D Fourier spectrum of the diffracted wavefront $g(x,y,z)|_{z=R}$ is given by a product of three components, that is, the spectrum on the surface of the hemisphere whose center is located at the origin and whose radius is $1/\lambda$ in the 3-D Fourier spectrum $F(u,v,w)$, the weight factor $\frac{i}{4\pi} \frac{1}{\sqrt{1/\lambda^2 - u^2 - v^2}}$, and the phase factor $\exp\left(i2\pi R\sqrt{1/\lambda^2 - u^2 - v^2}\right)$.

The diffraction distance R determines the phase components. In other words, the information of the diffracted wavefront from the 3-D object exists locally on the hemispherical surface in the 3-D Fourier space.

Next, the propagation direction of the diffracted wavefront is generalized, and the physical meaning of Eq. (6.71) is explained graphically based on Fig. 6.15.

Figure 6.15 shows that the 2-D Fourier spectrum of a diffracted wavefront is given by the spectrum on a hemisphere in the 3-D Fourier spectrum of the 3-D object. The rotational axis of the hemisphere is determined by a propagation direction. For example, in the case of the wavefront propagating in the $+z$ direction, the rotational axis becomes $+w$ direction. Similarly, the spectrum in the $+x_0$ direction corresponds to a hemisphere whose rotational axis is $+u$, and its spectrum is partially shared with that of diffraction in the $+z$ direction. Therefore, it is possible to express all of the diffracted wavefronts by the spectrum on a single sphere. All of the 3-D Fourier spectrum is never necessary. In other words, the spectrum on a single sphere includes all information on diffraction.

PROBLEMS

1. Consider a transparent object $g(x,y)$ located in between a positive lens (focal length f) and its focal point. The distance between the lens and the object is d. If the lens is illuminated with a monochromatic collimated light, find the amplitude of the wave at the focal plane of the lens.

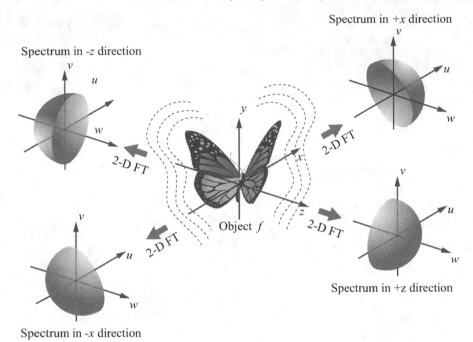

Figure 6.15 Fourier spectrum on the hemisphere, which corresponds to diffracted wavefronts in different directions.

2. Consider an object located at a distance $2f$ to the left of a positive lens (focal length f). Find the position and the magnification of the image of the object.

3. Two lenses have focal lengths f_1 and f_2.

   a. If the two thin lenses are placed in contact, determine the focal length of a lens combination.

   b. If the two lenses are placed at a distance of d, determine the focal length of a lens combination, where $d < f_1$ and $d < f_2$.

4. An object with the amplitude transmittance

$$t(x,y) = 1/2[1 + \cos 2\pi(x^2 + y^2)]$$

   is illuminated by monochromatic light. Determine the light amplitude at a distance d from the object.

5. Determine the lens transmittance as shown in Eq. (6.13) with the radius of curvature R_1 and R_2 and its thickness Δ_0.

6. Draw OTFs of an optical system with the pupil as shown in Fig. 6.16 for the cases of coherent and incoherent illumination. Consider the cases $D < a$, $D = a$ and $D > a$.

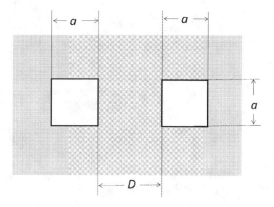

Figure 6.16   Double square pupile.

7. Discuss the image and the resolution limit of an object

$$t(x,y) = 1/2[1 + \cos(2\pi v_0 x)]$$

by an ideal optical system in the cases of coherent and incoherent illumination.

BIBLIOGRAPHY

J.W. Goodman. 1996. *Introduction to Fourier Optics*, 3rd ed. Roberts & Co. Englewood, CO.
O. K. Ersoy. 2006. *Diffraction, Fourier Optics and Imaging*. Wiley Interscience.
R. L. Easton, Jr. 2010. *Fourier Methods in Imaging*. John Wiley & Sons, Chichester, West Sussex.

REFERENCES

1. Lohmann, A. W., Mendlovic, D. and Zalevsky, Z. 1998. Fractional transforms in optics. *Prog Opt*. XXXVIII: 263–342.
2. Ozaktas, H. M., Zalevsky, Z. and Kutay, M. A. 2001. *The Spectrum for Omni-Directional Fractional Fourier Transform*. John Wiley & Sons, Chichester.
3. Sando, Y., Barada, D. and Yatagai, T. 2012. Fast calculation of computer-generated holograms based on 3-D Fourier spectrum for omnidirectional diffraction from a 3-D voxel-based object, *Opt Express*. 20: 20962–20969.
4. Kak, A. C. and Slaney, M. 1988. *Principles of Computerized Tomographic Imaging*. IEEE Press, 203–273.

7 Holography

Holography can record and reconstruct a complex amplitude from a 3-D object. In conventional photography, it is difficult to record directly the complex amplitude. In holography, recording and reconstructing of the complex amplitude are performed by superimposing a carrier wave to the complex amplitude to be recorded.

Holography is also able to realize and modify complex amplitude characteristics in spatial frequency filtering and therefore bring us a big progress in optical computing technology. Computer-generate hologram (CGH) is also discussed to generate complex amplitude filters with complexity and 3-D display of non-existing objects. Finally, a digital version of the holography, called the digital holography, is presented.

7.1 CONVENTIONAL OPTICAL HOLOGRAPHY

Consider recording the amplitude arriving from an object in the recording medium. This wave is called the object wave, whose complex amplitude is described by $f(x,y) = A(x,y)\exp[i\phi(x,y)]$, as shown in Fig. 7.1, where $A(x,y)$ is the amplitude of $f(x,y)$ and $\phi(x,y)$ its phase. If the object wave is recoded directly, its intensity is given by

$$I(x,y) = |A(x,y)\exp[i\phi(x,y)]|^2 = |A(x,y)|^2 \tag{7.1}$$

and therefore the phase information is lost.

Consider a plane wave tilted by an angle of θ to the x axis, which is called the reference wave,

$$r(x,y) = R\exp(i2\pi x \sin\theta/\lambda) \tag{7.2}$$

and by superimposing the reference wave to the object wave, the intensity at the recording plane is given by

$$\begin{aligned}
I(x,y) &= |f(x,y) + r(x,y)|^2 \\
&= |A(x,y)\exp[i\phi(x,y)] + R\exp(i2\pi x \sin\theta/\lambda)|^2 \\
&= |A(x,y)|^2 + |R|^2 \\
&\quad + A(x,y)R\exp\{i[\phi(x,y) - 2\pi x \sin\theta/\lambda]\} \\
&\quad + A(x,y)R\exp\{-i[\phi(x,y) - 2\pi x \sin\theta/\lambda]\} \\
&= |A(x,y)|^2 + |R|^2 + 2A(x,y)R\cos[\phi(x,y) - 2\pi x \sin\theta/\lambda].
\end{aligned} \tag{7.3}$$

This intensity includes the phase information of the object $\phi(x,y)$ as well as the amplitude $A(x,y)$. This technique is called holography, which records the complex amplitude of the object wave by using the reference wave. The third term in Eq. (7.3) is a cosine term, which represents interference fringes, and the recorded medium is called the hologram.

DOI: 10.1201/9781003121916-7

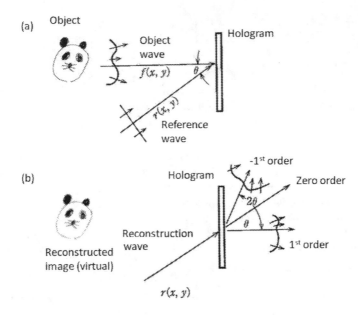

Figure 7.1 Principle of holography. (a) Recording of hologram and (b) reconstruction of hologram.

The amplitude transmittance of the hologram, the recorded intensity distribution, is given by

$$t(x,y) = t_0 + \gamma I(x,y). \tag{7.4}$$

If the hologram is illuminated by the same reference wave, the following waves are diffracted

$$t(x,y)r(x,y) = [t_0 + \gamma\{|A(x,y)|^2 + \gamma|R|^2]R\exp(i2\pi x \sin\theta/\lambda)$$
$$+ \gamma A(x,y)R^2 \exp[i\phi(x,y)]$$
$$+ \gamma A(x,y)R^2 \exp\{-i[\phi(x,y) - 4\pi x \sin\theta/\lambda]\}. \tag{7.5}$$

The first term in Eq. (7.5) is the zero-order diffraction wave, which propagates in the same direction as the reference wave and loses the phase information of the object. The second term is the +1st order diffraction wave, which reconstructs the object wave with a constant factor of γR^2 and propagates in the same direction as the original object wave. By looking into the hologram, the object image is observed at the same position as the object location. If the object is not plane but three-dimensional (3-D), the reconstructed image is 3-D. This is the main reason why the hologram is used for recording and displaying 3-D images. The third term is the -1st order diffraction wave, propagating to the 2θ direction. It consists of the original amplitude $A(x,y)$, but its phase is reversed, and therefore, the diffracted image is called the conjugate image.

7.2 COMPUTER GENERATED HOLOGRAPHY

The complex amplitude with the amplitude $A(x,y)$ and the phase $\phi(x,y)$ can be recorded and reconstructed by using a technique of holography was described in Section 7.1. Since the complex amplitude of an object should be interfered with a reference wave to record a hologram, the object should exist in the real world. It is very difficult to make a filter with complex characteristics by the holographic technique. The computer-generated hologram (CGH) can be synthesized by only computational procedures. This means that the wavefront to be recorded needs not always to exist in the real world. The technique of computer-generated holography is widely used to reconstruct non-existing objects, to synthesize spatial filters with complex characteristics, to generate wavefronts with ideal shapes which are difficult to be fabricated by conventional methods, and so on.

The procedure of synthesizing a CGH is shown in Fig. 7.2. At first, object image data are stored in a 2-D or 3-D model in a computer. Based on the diffraction theory, the complex amplitude arrived at a hologram plane from the object space is calculated. As mentioned in Chapter 6, the Fresnel diffraction and the Fraunhofer diffraction are calculated by using the Fast Fourier Transform (FFT). Examples of the procedures of the diffraction calculation are described in Appendix A. Finally, the calculated complex amplitude of the hologram is drawn in some formats. Cell-oriented and point-oriented CGHs are commonly used hologram formats.

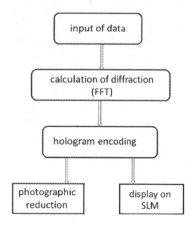

Figure 7.2 Procedure of synthesizing of computer-generated hologram.

7.2.1 CELL-ORIENTED CGH

One of the typical cell-oriented CGHs is the Lohmann CGH. The Lohmann CGH is a binary hologram whose amplitude transmittance is binary, 0 or 1 [2]. Consider the complex amplitude $H(v_x, v_y)$ at a sampling point (m,n)

$$H_{mn} = H(m\Delta v, n\Delta v) = A_{mn}\exp(i\phi_{mn}), \tag{7.6}$$

where the spacing of sampling points is denoted by Δv and the amplitude and the phase of H_{mn} denoted by A_{mn} and ϕ_{mn}, respectively.

The cell structure of the Lohmann CGH is shown in Fig. 7.3.

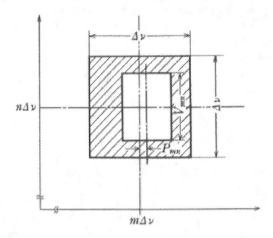

Figure 7.3 Cell structure of Lohmann hologram.

The Lohmann hologram consists of sampling cells with spacing Δv. Each cell includes a rectangular aperture with the height V_{mn} and the shift P_{mn} from the center of the aperture so as to realize the amplitude A_{mn} and the phase ϕ_{mn} at the sampling point (m,n). The amplitude transmittance of the Lohmann CGH is given by

$$t(v_x, v_y) = \sum_m \sum_n \text{rect}\left(\frac{v_x - m\Delta v - P_{mn}}{c}\right) \times \text{rect}\left(\frac{v_y - n\Delta v}{V_{mn}}\right), \tag{7.7}$$

where c denotes the width of the rectangular aperture. If the aperture at the sampling point (m,n) is illuminated by a plane wave $\exp(-i2\pi x_0 v_x)$. The diffracted wave from the aperture is given by

$$\iint \text{rect}\left(\frac{v_x - m\Delta v - P_{mn}}{c}\right) \times \text{rect}\left(\frac{v_y - n\Delta v}{V_{mn}}\right) \exp(-i2\pi x_0 v_x) dv_x dv_y \tag{7.8}$$
$$= cV_{mn}\text{sinc}(cx_0)\exp[-i2\pi x_0(m\Delta v + P_{mn})].$$

Since the amplitude of this diffracted wave should be equal to H_{mn}, we have

$$A_{mn} = cV_{mn}\text{sinc}(cx_0) \tag{7.9}$$

$$\phi_{mn} = -2\pi x_0(m\Delta v + P_{mn}). \tag{7.10}$$

If we use the N-th diffracted wave,

$$x_0\Delta v = N. \tag{7.11}$$

Usually, we set $N = 1$ because we use the first-order diffracted wave. To increase the diffraction efficiency, we set $c = \Delta v/2$. Finally, we have a simpler version of the height of the aperture and its shift from Eqs. (7.9) and (7.10)

$$V_{mn} = \frac{A_{mn}\Delta v}{\max(A_{mn})} \tag{7.12}$$

$$P_{mn} = \frac{\phi_{mn}\Delta v}{2\pi}, \tag{7.13}$$

where $\max(A_{mn})$ is the maximum of A_{mn}. Figure 7.4 shows an example of the Lohmann CGH (a) and its reconstructed image (b). The sampling number is 256 × 256.

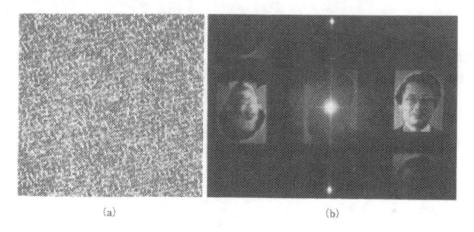

(a)   (b)

Figure 7.4   Lohmann hologram (a) and its reconstructed image (b).

7.2.2   POINT-ORIENTED CGH

For simplicity, consider a Fourier transform CGH which reconstructs an image $f(x,y)$. In the point-oriented CGH, it is noted that the transmittance of the hologram $T(v_x, v_y)$ is real and positive. Therefore its Fourier transform $t(x,y)$ should satisfy

$$t(x,y) = t^*(-x,-y). \tag{7.14}$$

If we set

$$t(x,y) = f(x - x_0, y - y_0) + f^*(-x - x_0, -y - y_0) \tag{7.15}$$

to reconstruct the image $f(x,y)$, its Fourier transform $T(v_x, v_y)$ gives the transmittance of the hologram, as shown in Fig. 7.5.

In the case when the wave to be reconstructed is described by a simple function $\phi(x,y)$, the fringe pattern of the CGH with the carrier frequency α can be written as

$$h(x,y) = A + B\cos[2\pi\alpha x + \phi(x,y)]. \tag{7.16}$$

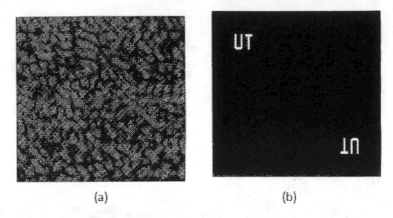

(a)   (b)

Figure 7.5   Point-oriented CGH (a) and its reconstructed image (b).

If the fringe contrast B is assumed to be constant, the transmittance of the binary CGH is given by

$$t(x,y) = \begin{cases} 1 & : \frac{1}{2} + \frac{1}{2}\cos[2\pi\alpha x + \phi(x,y)] > c \\ 0 & : \text{otherwise,} \end{cases} \tag{7.17}$$

where c is a constant $(0 < c < 1)$ and usually $c = 1/2$. An example of the fringe-type CGH is shown in Fig. 7.6.

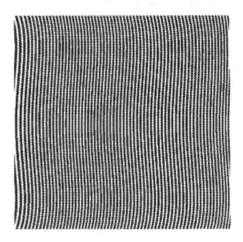

Figure 7.6   Fringe-type CGH.

7.2.3 KINOFORM

In the holography, the reference wave is used to introduce a carrier frequency into the phase of the hologram. The carrier frequency enables us to separate the zero and ± 1st order diffracted waves in reconstruction. If we can generate or display the complex amplitude distribution directly, the carrier frequency is not necessary. For simplicity, the amplitude of the diffracted wave from a diffuse object is considered to be constant and its phase to be modulated. In such a case, the phase-only optical element generated by a computer is called a kinoform [2], whose diffraction efficiency is higher than the conventional CGHs. The phase information of the kinoform is represented by a relief profile on a recording material such as a photographic emulsion or is written on a spatial light modulator. The kinoform can include the reconstruction noise caused by the amplitude variation neglect and the phase quantization. Because the noise in the reconstructed image is a serious problem, it is necessary that the phase distribution of the kinoform should be optimized to decrease the noise. Several methods have been discussed. Figure 7.7 shows a result of a simulated annealing iteration method [3]. The reconstructed noise is reduced due to increase of the iteration numbers.

Figure 7.7 Reconstructed images of the phase optimized kinoform by an iteration method. Number of iteration is (a) 10, (b) 100, (c) 300, and (d) 500.

7.3 DIGITAL HOLOGRAPHY

A technique of recording holograms in a digital memory is called digital holography. Figure 7.8 shows an optical setup of the digital holography. Similar to the conventional holography, the intensity of the interference fringe of the object wave and the reference wave is given by Eq. (7.3). Since the pixel number of an image sensor is limited, the carrier frequency should be suppressed as low as possible. Accordingly, the angle between the object and the reference wave is set as small as possible and the size of the object is limited.

Consider the case when the reconstructed image of the object is located at the original place of the object located, as shown in Fig. 7.1. The reconstructed wave is the conjugate wave of the reference wave in this case. The reconstruction wave is the wave propagating backward from the hologram plane $r^*(x,y) = R\exp(-i2\pi x \sin\theta/\lambda)$. The diffracted wave from the hologram is given by

$$t(x,y)r^*(x,y) = [t_0 + \gamma|A(x,y)|^2 + \gamma R^2]R\exp\left(-i\frac{2\pi x}{\lambda}\sin\theta\right)$$
$$+\gamma A(x,y)R^2\left[i\phi(x,y) - i\frac{4\pi x}{\lambda}\sin\theta\right]$$
$$+\gamma A(x,y)R^2\exp[-i\phi(x,y)]. \tag{7.18}$$

The third-term gives the real image located at the original position. The reconstructed image is calculated numerically by diffraction formulae back to the image plane from the hologram plane. Similar to CGHs, the diffraction calculation is performed by the Fresnel diffraction, the Fraunhofer diffraction or the angular spectrum method. In this numerical calculation, the position of the object should be known precisely. An example of numerical procedures for the digital holography is given in Appendix B.

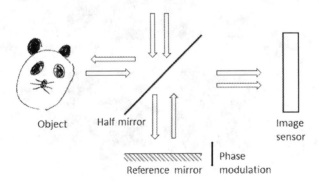

Figure 7.8 Optical setup for digital holography.

Next, consider the in-line holography. To eliminate the zero diffraction wave and to solve the twin image problem in the in-line hologram, the phase modulation method is used. The intensity of the in-line hologram is described by Eq. (7.3) with $\theta = 0$, but the phase modulation of δ is introduced.

$$I(x,y,\delta) = |A(x,y)|^2 + |R|^2 + 2A(x,y)R\cos[\phi(x,y) - \delta]. \tag{7.19}$$

If we record four holograms, $I(x,y,0), I(x,y,\pi/2), I(x,y,\pi)$, and $I(x,y,3\pi/2)$ with different reference phases, $\delta = 0, \pi/2, \pi$ and $3\pi/3$, respectively, the amplitude of the object wave on the image sensor is given by

$$A(x,y)\exp[i\phi(x,y)] = \frac{1}{4R}\{[I(x,y,0) - I(x,y,\pi)] + i[I(x,y,\pi/2) - I(x,y,3\pi/2)]\}.$$

$$(7.20)$$

The Fresnel diffraction of the object wave $A(x,y)\exp[i\phi(x,y)]$ to the object plane gives the reconstructed image.

PROBLEMS

1. Describe the features and applications of holography.
2. Describe practical applications of computer-generated holography.

BIBLIOGRAPHY

Goodman, J. W. 1996. *Introduction to Fourier Optics*. 3rd ed. Roberts & Co., Englewood, CO.
Hariharan, P. 1984. *Optical Holography*. Cambridge Univ. Press, Cambridge, UK.
Schnars, U. and Jueptner, W. 2005. *Digital Holography*. Springer, Berlin.

REFERENCES

1. Lohmann, A. W. and Paris, D. P. 1967. Binary Fraunhofer holograms generated by computer. *Appl. Opt.* 6: 1739.
2. Lesem, L. P., Hirsch, P. M. and Jordan, Jr., J. A. 1969. The kinoform: a new wave front reconstruction device. *IBM J. Res. Dev.* 13: 150.
3. Yoshikawa, N. and Yatagai, T., 1994. Phase optimization of a kinoform by simulated annealing. *Appl. Opt.* 33: 863.

8 Optical Computing

As an important application of Fourier optics, consider optical computing, a technology of optical information processing using parallel and high-speed of information propagation of light. This technology includes analog image processing and digital parallel processing. In this chapter, parallel analog optical computing is discussed, which is mostly related to Fourier optics.

Two-dimensional (2-D) spectrum is obtained by optical Fourier transforming input signal and then modified by spatial frequency filter. This technique is called spatial frequency filtering, which is one of the most typical parallel optical computing operations. This operation can be considered to be an optical convolution operation.

8.1 SPATIAL FREQUENCY FILTERING

Spatial frequency filtering is defined as spatial frequency modification in order to enhance a part of the input signal, to extract a specific component of the signal or detect the signal in noise. The Fourier transforming ability of a lens makes Fourier transform of an input image and this enables us to perform spatial frequency filtering for 2-D images. This is one of the fundamental techniques in optical computing [1, 2].

Consider a coherent optical system as shown in Fig. 8.1. An input image $f(x,y)$ located in the plane P_1 is illuminated by a collimated coherent light and is Fourier transformed by the lens L_1 and the Fourier spectrum

$$F(v_x, v_y) = \mathscr{F}[f(x,y)] \tag{8.1}$$

is obtained in plane P_2. A filter whose complex amplitude transmittance is $H(v_x, v_y)$ is set in plane P_2. The complex amplitude just after the plane P_2 is $F(v_x, v_y) \cdot H(v_x, v_y)$, which means that the spectrum $F(v_x, v_y)$ of the input is modified by $H(v_x, v_y)$. Fourier transforming $F(v_x, v_y) \cdot H(v_x, v_y)$ by the lens L_2 gives

$$\mathscr{F}[F(v_x, v_y) \cdot H(v_x, v_y)] = h * f(x_0, y_0) \tag{8.2}$$

in the output plane P_3, where

$$h(x_0, y_0) = \mathscr{F}[H(v_x, v_y)] \tag{8.3}$$

and $h(x_0, y_0)$ is the impulse response of the filter $H(v_x, v_y)$. The optical system shown in Fig. 8.1 calculates the convolution between the input $f(x,y)$ and the impulse response $h(x_0.y_0)$ of the filter. Since the optical system shown in Fig. 8.1 performs Fourier transform two times, this optical system is called the double diffraction optical system. Since Fourier transform without a phase component is obtained in the case when the input locates at a distance of the focal length f of the lens, as noted in Sec. 6.2, the optical system shown in Fig. 8.1 is also called "4-f optical system."

DOI: 10.1201/9781003121916-8

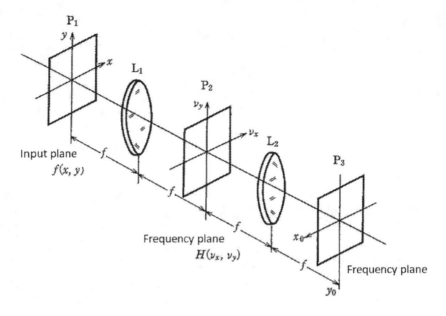

Figure 8.1   Optical system of spatial filtering.

Figure 8.2 shows Fourier spectrum patterns for alphabets and numbers. The Fourier spectrum patterns show characteristics of input patterns, like shape, width and direction of components, and so on. Sometimes, only by the shape of the spectrum pattern recognition or classification is performed.

8.1.1   LOW-PASS AND HIGH-PASS FILTERS

In general, the lower frequency component includes the rough structural information of the image, and the edge or fine structural information contributes to the higher-frequency component. If the lower frequency component is passed by a filter, the noise including the higher frequency component can be reduced. Such filter is called the low-pass filter. On the other hand, the filter which passes the higher frequency component is called the high-pass filter. The high-pass filter is used to detect image boundary and enhance fine structure of image. The band-pass filter, which passes or cuts specific spectrum component, is used to reduce half-tone dot meshing structure, for example. Figure 8.3 the frequency responses for a low-pass filter (a) and a high-pass filter (b).

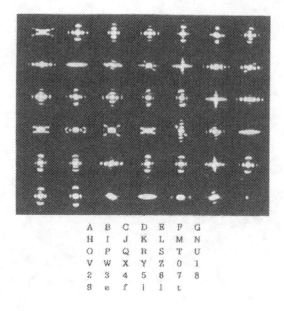

Figure 8.2   Fourier spectrum of alphabets and numbers.

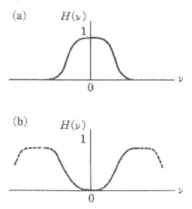

Figure 8.3   Frequency response of (a) low-pass filter and (b) high-pass filter.

8.1.2   DIFFERENTIATION AND LAPLACIAN FILTERS

The differential filter is used for the same purpose of the high-pass filter. Consider a complex amplitude of an input image $f(x,y)$ given by

$$f(x,y) = \iint_{-\infty}^{\infty} F(v_x, v_y) \exp[i2\pi(v_x x + v_y y)] dv_x dv_y. \qquad (8.4)$$

Its differential to the x-direction is

$$\frac{\partial f}{\partial x} = \iint (i2\pi v_x)F(v_x, v_y)\exp[i2\pi(v_x x + v_y y)]dv_x dv_y. \tag{8.5}$$

Therefore, the frequency response of the differential filter is given by

$$H(v_x, v_y) = i2\pi v_x \tag{8.6}$$

as shown in Fig. 8.4(a). Since

$$\left(\frac{\partial^2 f}{\partial x^2} + \frac{\partial^2 f}{\partial x^2}\right) = -\iint (4\pi^2)(v_x^2 + v_y^2)F(v_x, v_y)\exp[i2\pi(v_x x + v_y y)]dv_x dv_y, \tag{8.7}$$

the frequency response of Laplacian filter is given by

$$H(v_x, v_y) = -4\pi^2(v_x^2 + v_y^2) \tag{8.8}$$

as shown in Fig. 8.4(b) [3]. Since the differential filters are a kind of high-pass filters, it should be noted that the noise components are enhanced in some cases.

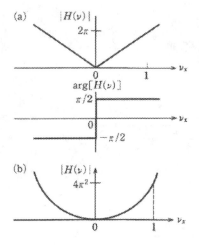

Figure 8.4 Frequency response of (a) differential filter and (b) Laplacian filter.

8.1.3   PHASE-CONTRAST FILTER

The phase contrast filter, which transfers the phase distribution to the intensity, is used to image the phase object, e.g., semi-transparent biological samples. The amplitude distribution of a transparent sample is given by

$$f(x,y) = A\exp[i\phi(x,y)], \tag{8.9}$$

where A is a constant amplitude and $\phi(x,y)$ a phase distribution of the sample. If the sample image is detected as the intensity, we have

$$I(x,y) = A^2 \tag{8.10}$$

and no contrast of the image is obtained. Consider the case when its phase change is very small, $\phi \ll 1$, the object amplitude is given by

$$f(x,y) = A[1 + i\phi(x,y)]. \tag{8.11}$$

Since the amplitude A is constant, the spectrum due to the object amplitude concentrates near the zero frequency, but the spectrum of the phase $\phi(x,y)$ distributes widely. If the phase filter which changes the higher frequency component by $\pi/2$ is used, the output intensity is

$$I(x,y) = |A[1 - \phi(x,y)]|^2 \tag{8.12}$$
$$\fallingdotseq A^2 - 2A^2\phi(x,y)$$

and therefore, the phase information $\phi(x,y)$ is changed to the contrast. This filter is called the phase contrast filter [4] and is utilized in some biological microscopic objective lenses.

8.1.4 SUPER RESOLUTION AND APODIZATION

The resolution of the optical system is defined as the reciprocal of the Airy disc radius according to the Rayleigh criterion, as described in Sec. 6.6. This means that the resolution is increased by reducing the Airy disc. Consider the perfect imaging system whose pupil function is circular with the diameter D. If the circular aperture stops with the diameter $\varepsilon D(0 < \varepsilon < 1)$ is set in the center of the pupil, as shown in Fig. 8.5, the amplitude of the point image is given by

$$u(w) = \pi A' \left(\frac{D}{2}\right)^2 \cdot \frac{2J_1\left(\frac{kD}{2R}w\right)}{\frac{kD}{2R}w} - \pi A' \left(\frac{\varepsilon D}{2}\right)^2 \cdot \frac{2J_1\left(\frac{k\varepsilon D}{2R}w\right)}{\frac{k\varepsilon D}{2R}w} \tag{8.13}$$

according to Eq. (2.66). Its intensity is

$$I(w) = |u(w)|^2 = I_0 \left| \frac{2J_1\left(\frac{kD}{2R}w\right)}{\frac{kD}{2R}w} - \varepsilon^2 \frac{2J_1\left(\frac{k\varepsilon D}{2R}w\right)}{\frac{k\varepsilon D}{2R}w} \right|^2, \tag{8.14}$$

where

$$I_0 = \left[\pi A' \left(\frac{D}{2}\right)^2\right]^2. \tag{8.15}$$

As shown in Fig. 8.5, a narrower diffraction pattern than the perfect optical system is obtained. The method which changes the pupil function to improve the resolution higher than the perfect optical system is called the super resolution. It should be noted that the width of the main peak is reduced, but the intensity of the outer annular peaks is increased in the present super resolution [5, 6].

In a converse method, the outer area of the pupil function is reduced gradually so that the intensity of the outer annular peaks is reduced, but the width of the main peak is increased. This method is called apodization, which is used in imaging of a small dark object near a bright object.

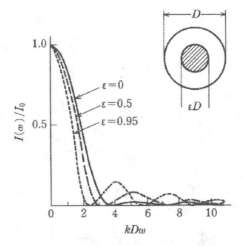

Figure 8.5   Diffraction of annular aperture.

8.2   MATCHED FILTER

The signal to noise ratio (SNR) is defined as the measure of signal detection performance:

$$\text{SNR} = \frac{\text{Signal power}}{\text{Noise power}}. \tag{8.16}$$

In order to detect a specific pattern and its position, the filter which maximizes SNR of the output is designed. This is called the matched filter [7].

In pattern recognition, one of the main issues is to search for the positions where known target patterns are in the input image. Since the problem is only to determine whether the patterns to be detected exist or not, the output does not always retain the original shape. Consider a noise reduction method by using an appropriate filter for the input image with the known pattern and the noise so as to make pattern recognition with ease [7, 8].

Consider the input $g(x, y)$ consisting of the signal $f(x, y)$ with the noise $n(x, y)$.

$$g(x, y) = f(x, y) + n(x, y). \tag{8.17}$$

It should be noted that the noise $n(x, y)$ is additive to the signal $f(x, y)$. Let the response function of the matched filter be $h(x, y)$, as shown in Fig. 8.6.

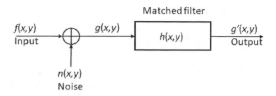

Figure 8.6 Matched filter.

The matched filter output $g'(x,y)$ for the signal $f(x,y)$ is given by

$$g'(x,y) = f * h(x,y) \tag{8.18}$$

and for the noise

$$n'(x,y) = n * h(x,y) \tag{8.19}$$

If the noise $n(x,y)$ is the white noise, the spectrum of the noise is constant. This means that the noise $n(x,y)$ is described by a random variable. The power of the noise is simply given by $|n(x,y)|^2$, which is also a random variable and nondeterministic. Its ensemble average $E[|n'(x,y)|^2]$ gives a deterministic value. The signal to noise ratio, SNR, for the matched filter is defined as

$$\text{SNR} = \frac{|g'(x,y)|^2}{E[|n'(x,y)|^2]}. \tag{8.20}$$

The ensemble average of the noise $E[|n'(x,y)|^2]$ is rewritten as

$$E[|n'(x,y)|^2] = N^2 \iint |H(v_x,v_y)|^2 dv_x dv_y, \tag{8.21}$$

where N^2 denotes the power spectrum of the noise, and $H(v_x,v_y)$ is the Fourier transform of $h(x,y)$, which is the frequency response of the matched filter. Applying the convolution theorem (3.89) to Eq. (8.18) and using Eq.(8.21), we have

$$\text{SNR} = \frac{\left| \iint F(v_x,v_y) \cdot H(v_x,v_y) \exp[i2\pi(v_x x + v_y y)] dv_x dv_y \right|^2}{N^2 \iint |H(v_x,v_y)|^2 dv_x dv_y}. \tag{8.22}$$

By using Schwarz' inequality

$$\left| \iint a(x,y) b(x,y) dx dy \right|^2 \leq \iint |a(x,y)|^2 dx dy \cdot \iint |b(x,y)|^2 dx dy \tag{8.23}$$

with equality if $a = b^*$, we have

$$\text{SNR} \leq \frac{1}{N^2} \iint |F(v_x,v_y)|^2 dv_x v_y. \tag{8.24}$$

The maximum of SNR is obtained when

$$H(v_x, v_y) = F^*(v_x, v_y) \exp[-i2\pi(v_x x_0 + v_y y_0)]. \tag{8.25}$$

This is the frequency response of the matched filter, which is the complex conjugate of the Fourier transform of the pattern to be detected.

Since the output of the matched filter is given by $f * h$, from Eq. (8.18), we have

$$\begin{aligned}
g'(x, y) &= \iint H(v_x, v_y) \cdot F(v_x, v_y) \exp[i2\pi(v_x x + v_y y)] dv_x dv_y \\
&= \iint F^*(v_x, v_y) \exp[-i2\pi(v_x x_0 + v_y y_0)] \\
&\quad \times F(v_x, v_y) \exp[i2\pi(v_x x + v_y y)] dv_x dv_y \\
&\quad + \iint N(v_x, v_y) \cdot F^*(v_x, v_y) \exp[-i2\pi(v_x x_0 + v_y y_0)] \\
&\quad \times \exp[i2\pi(v_x x + v_y y)] dv_x dv_y \\
&= f \star f(x - x_0, y - y_0) + n \star f(x - x_0, y - y_0). \tag{8.26}
\end{aligned}$$

The peak of the autocorrelation is located at the position (x_0, y_0) of the pattern to be detected. Since the peak of the autocorrelation is much larger than the correlation between the pattern to be detected and the noise, the location of the pattern to be detected is very easily recognized.

Consider the matched filter using a holographic technique to detect a signal pattern $f(x, y)$. It should be noted that the -1st order diffracted wave from the Fourier transform of $f(x, y)$ gives the frequency response of Eq. (8.25). An optical setup for matched filtering is shown in Fig. 8.7. A pattern $f(x, y)$ to be detected is located in P_1 plane, and the Fourier transform $F(v_x, v_y)$ is formed by a Fourier transform lens L_1 in P_2 plane. Using reference wave $R \exp(i2\pi v_x f_L \sin\theta)$, a Fourier transform hologram is formed in P_2 plane, where f_L denotes the focal length of the lens L_2.[1] As same as Eqs. (7.3) and (7.5), the amplitude transmittance of the hologram is given as

$$\begin{aligned}
t(v_x, v_y) &= t_1 + \gamma R F(v_x, v_y) \exp(-i2\pi v_x f_L \sin\theta) \\
&\quad + \gamma R F^*(v_x, v_y) \exp(i2\pi v_x f_L \sin\theta), \tag{8.27}
\end{aligned}$$

where

$$t_1 = t_0 + \gamma(|F|^2 + |R|^2). \tag{8.28}$$

The desired term of the matched filter is the third term of Eq. (8.27).

To make matched filtering, an input pattern is located in the P_1 plane. The wave passing through the matched filter is Fourier transformed and then the output image is obtained in the P_3 plane. In the output plane as shown in Fig. 8.7, the zero-order diffracted wave locates in the central area and the ± 1st order diffracted waves locate

---

[1]The plane wave tilted by an angle of θ to the filter plane is represented by $\exp(i2\pi x_0 \sin\theta/\lambda)$, where x_0 denotes the coordinate in the filter plane. Because the spatial frequency v_x in the filter plane is given by $v_x = x_0/\lambda f_L$, the tilted plane wave is represented by $\exp(i2\pi v_x f_L \sin\theta)$.

the right and the left sides corresponding to the convolution and the correlation terms, $g * f(x-d,y)$ and $g \star f(x+d,y)$, respectively, where

$$d = f_L \sin\theta. \tag{8.29}$$

Figure 8.8 shows an example of a matched filter (a) for detecting the character **E**, the impulse response (b) and the correlation output (c). It should be noted that the bright correlation spot locates at the position where the character **E** exists.

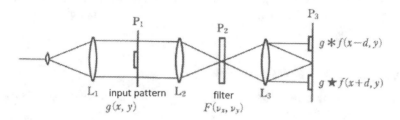

Figure 8.7 Optical setup for matched filtering.

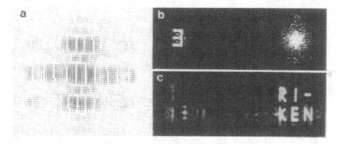

Figure 8.8 Output of matched filtering. (a) Matched filter computer generated, (b) impulse response of (a), and (c) correlation output.

8.3 OPTIMUM FILTER FOR ADDITIVE NOISE

The matched filter can detect the position of a target image in an input pattern with random noise. Here, a filter that can suppress additive noise and restore the degraded image is discussed. Consider an optical system shown in Fig. 8.9. The output of the imaging system $g(x,y)$ with additive noise $n(x,y)$ is given by

$$g(x,y) = f(x,y) * h(x,y) + n(x,y), \tag{8.30}$$

where $f(x,y)$ and $h(x,y)$ denote the input image and the point spread function of the system, respectively. Consider the noise be a wide-sense stationary random process with known second-order statistics and the noise be statistically independent of the image $g(x,y)$.

An optimum filter is designed, which can minimize the noise and the image degradation. The image after the optimum filter is given by

$$g'(x,y) = g(x,y) * t(x,y), \tag{8.31}$$

where $t(x,y)$ denotes the point spread function of the restoration system. The optimum filter minimizes the mean of the square error, that is, the average of the square difference between the original input $f(x,y)$ and the filter output $g'(x,y)$

$$E = \langle |f(x,y) - g'(x,y)|^2 \rangle, \tag{8.32}$$

where $\langle \cdots \rangle$ denotes ensemble average.

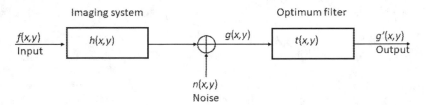

Figure 8.9 Optimum filter for additive noise.

Its solution is given by

$$t(x,y) * \phi_{gg}(x,y) = \phi_{fg}(x,y), \tag{8.33}$$

where ϕ_{gg} and ϕ_{fg} denote the auto-correlation of $g(x,y)$ and the cross-correlation of $f(x,y)$ and $g(x,y)$ [9]. Fourier transforming both term of Eq. (8.33) gives

$$T(v_x, v_y)\Phi_{gg}(v_x, v_y) = \Phi_{fg}(v_x, v_y), \tag{8.34}$$

where $T(v_x, v_y)$, $\Phi_{gg}(v_x, v_y)$ and $\Phi_{fg}(v_x, v_y)$ denote the Fourier transforms of $t(x,y)$, $\phi_{gg}(x,y)$ and $\phi_{fg}(x,y)$, respectively. Furthermore, $\Phi_{gg}(v_x, v_y)$ and $\Phi_{fg}(v_x, v_y)$ are the power spectrum of $g(x,y)$ and the cross power spectrum of $f(x,y)$ and $g(x,y)$. Then the optimum filter is given by

$$T(v_x, v_y) = \frac{\Phi_{fg}(v_x, v_y)}{\Phi_{gg}(v_x, v_y)} = \frac{|F(v_x, v_y)|^2 H^*(v_x, v_y)}{|F(v_x, v_y)|^2 |H(v_x, v_y)|^2 + |N(v_x, v_y)|^2}, \tag{8.35}$$

where $N(v_x, v_y)$ denotes the Fourier spectrum of the noise $n(x,y)$.

If the imaging system is ideal and therefore $H(v_x, v_y) = 1$, Eq. (8.35) is represented by

$$T(v_x, v_y) = \frac{|F(v_x, v_y)|^2}{|F(v_x, v_y)|^2 + |N(v_x, v_y)|^2}. \tag{8.36}$$

This is called the Wiener filter.

8.4 OPTIMUM FILTER FOR MULTIPLICATIVE NOISE

Spatial frequency filters for additive noise, such as the matched filter and the Wiener filter, have been discussed previously. Now, consider the noise, which is not additive but multiplicative, such as speckle noise. The intensity of the speckle noise depends on the intensity of the signal.

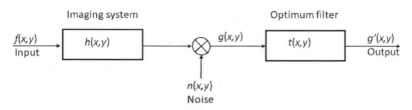

Figure 8.10 Optimum filter for multiplicative noise.

A model of a restoring spatial filter for degraded patterns with the multiplicative noise is shown in Fig. 8.10 [10]. The degraded image is represented by

$$g(x,y) = [f(x,y) * h(x,y)] \cdot n(x,y). \tag{8.37}$$

The output $g'(x,y)$ of the restoring filter $t(x,y)$ is given by

$$g'(x,y) = g(x,y) * t(x,y). \tag{8.38}$$

The problem is to minimize the mean of the square error defined in Eq. (8.32). As the same as the case of additive noise, the frequency response of the optimum filter is given by

$$T(v_x, v_y) = \frac{\Phi_{fg}(v_x, v_y)}{\Phi_{gg}(v_x, v_y)}, \tag{8.39}$$

where $\Phi_{fg}(v_x, v_y)$ is the cross power spectrum of $f(x,y)$ and $g(x,y)$,

$$\Phi_{fg}(v_x, v_y) = \langle n \rangle F^*(v_x, v_y) \Phi_{ff}(v_x, v_y) \tag{8.40}$$

and $\Phi_{gg}(v_x, v_y)$ is the power spectrum of $g(x,y)$,

$$\Phi_{gg}(v_x, v_y) = |F(v_x, v_y)|^2 \Phi_{ff}(v_x, v_y) * \Phi_{nn}(v_x, v_y). \tag{8.41}$$

Finally, the optimum filter for the multiplicative noise is given by

$$T(v_x, v_y) = \frac{\langle n \rangle F^*(v_x, v_y) \Phi_{ff}(v_x, v_y)}{|F(v_x, v_y)|^2 \Phi_{ff}(v_x, v_y) * \Phi_{nn}(v_x, v_y)}, \tag{8.42}$$

where $\langle n \rangle$ denotes the mean of the noise and $\Phi_{nn}(v_x, v_y)$ the power spectrum of the noise.

Next, consider the case

$$n(x,y) = \langle n \rangle + m(x,y), \tag{8.43}$$

where $m(x, y)$ is random noise with zero mean. In this case, the power spectrum of the noise is given by

$$\Phi_{nn}(v_x, v_y) = \langle n \rangle^2 \delta(v_x, v_y) + \Phi_{mm}(v_x, v_y), \tag{8.44}$$

where $\Phi_{mm}(v_x, v_y)$ denotes the power spectrum of $m(x, y)$. Then Eq. (8.42) is

$$T(v_x, v_y) = \frac{\langle n \rangle F^*(v_x, v_y) \Phi_{ff}(v_x, v_y)}{\langle n \rangle^2 |F(v_x, v_y)|^2 \Phi_{ff}(v_x, v_y) + |F(v_x, v_y)|^2 \Phi_{ff}(v_x, v_y) * \Phi_{mm}(v_x, v_y)}. \tag{8.45}$$

It should be noted that when the power spectrum $\Phi_{ff}(v_x, v_y)$ is small compared to $\langle n \rangle^2$, the optimum filter reduces to the inverse filter,

$$T(v_x, v_y) \propto \frac{1}{F(v_x, v_y)}. \tag{8.46}$$

When the width of the significant region of $\Phi_{ff}(v_x, v_y)$ is smaller than that of $\Phi_{mm}(v_x, v_y)$, the second term of the denominator of Eq. (8.45) is approximated by $\langle f \rangle^2 \Phi_{mm}(v_x, v_y)$, where $\langle f \rangle$ denotes the mean of $f(x, y)$. In this case, the optimum image restoration filter for the multiplicative noise is given by

$$T(v_x, v_y) = \frac{\langle n \rangle F^*(v_x, v_y) \Phi_{ff}(v_x, v_y)}{\langle n \rangle^2 |F(v_x, v_y)|^2 \Phi_{ff}(v_x, v_y) + \langle f \rangle^2 \Phi_{mm}(v_x, v_y)} \tag{8.47}$$

which is very similar to the optimum filter for the additive noise derived [11].

To verify the theoretical results, a numerical simulation was performed. For simplicity, we present a 1-D and ideal imaging case, that is, $H(v_x, v_y) = 1$. Consider that the original image $f(x, y)$ consists of two rectangular functions and dark background

$$f(x, y) = \text{rect}\left(\frac{x - a}{a}\right) + \text{rect}\left(\frac{x + a}{a}\right) + b, \tag{8.48}$$

where a and b denote constants. The noise $n(x, y)$ is generated by pseudo-random number generator in a computer. Figure 8.11(a) and (b) show the original image $f(x)$ described by Eq. (8.48) with $a = 1.0$ and $b = 0.1$, and its noisy version $g(x)$, respectively. After filtering with the optimum filter of Eq. (8.42) shown in (c), we obtained the filtered image of (d).

8.5 SPECTRUM ANALYZER

Since the Fourier transform effect of a lens enables us to compute the Fourier spectrum of a signal $f(x, y)$, 2-D Fourier transform can be used in the power spectrum measurement of 2-D patterns, the diameter measurement of small particles, and so on [12]. An equipment for analyzing the spectrum of the time-dependent signal $f(t)$ is called the spectrum analyzer. To compute the spectrum of the time-dependent signal $f(t)$ by optical means, the time-dependent signal should be transformed to the

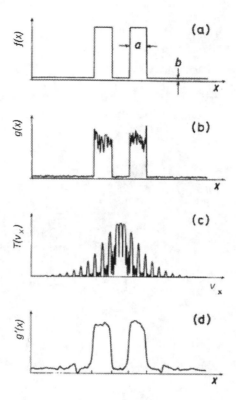

Figure 8.11 Numerical experiment of the optimum filter for the multiplicative noise. (a) Original image, (b) image degraded by multiplicative noise, (c) transfer function of the optimum filter, and (d) filtered image. The curves in this figure are normal sized with these maximum values.

spatial signal $f(x)$. The spatial light modulator (SLM) is used in this purpose [13]. For example, as shown in Fig. 8.12, the time dependent signal

$$s(t) = f(t)\cos(2\pi v_a t) \qquad (8.49)$$

is input to an ultra-sound spatial light modulator, where v_a denotes the carrier frequency of the acoustic wave. This signal propagating in the ultra-sound SLM generates the refractive index modulation proportional to $s(x - vt)$. The optical wavefront after passing through the SLM is written as

$$\exp[i\alpha s(x - vt)] \approx 1 + i\alpha s(x - vt) = 1 + i\alpha f(x - vt)\cos[k_a(x - vt)], \qquad (8.50)$$

when the refractive index change is small, where α is a constant and k_a the wave number of the ultra sound wave. The diffracted wave at the focal plane of a lens is given by

$$\mathscr{F}\{1 + i\alpha f(x - vt)\cos[k_a(x - vt)]\}$$
$$= \delta(v_x) + i\alpha\mathscr{F}[f(x)\cos(k_a x)]\exp(-i2\pi v_x vt). \qquad (8.51)$$

The amplitude of the first-order diffracted wave is $F(v_x)\exp(-i2\pi v_x vt)$ and is detected by a 1-D photo-detector array so that the power spectrum $|F(v_x)|^2$ of the signal $f(x)$ is obtained.

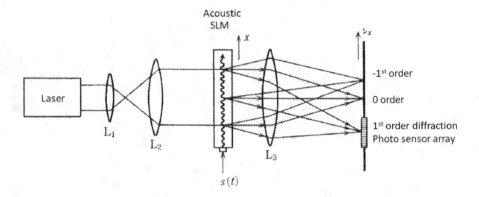

Figure 8.12   Spectrum analyzer.

8.6   OPTICAL CORRELATOR

Two-D optical correlation is performed by using spatial frequency, as described in Sec. 8.1 on spatial frequency filtering. Here, one-D optical correlator is described [14]. To calculate the correlation of the time-depending signal, the time-dependent signal should be transformed into the optical signal by using an acoustic SLM. The correlation function of two input signals $f_1(t)$ and $f_2(t)$ is defined as

$$\Phi_{12}(t) = \int_{-\infty}^{\infty} f_1(\tau)f_2(\tau - t)d\tau \qquad (8.52)$$

or

$$\Phi_{12}(\tau) = \int_{-\infty}^{\infty} f_1(t)f_2(t - \tau)dt. \qquad (8.53)$$

The correlation defined by Eq. (8.52) is called the space integral type because the integral variable τ is a spatial variable x. On the other hand, the correlation defined by Eq. (8.53) is called the time integral type because the integral variable t is a time variable t.

8.6.1 SPACE-INTEGRAL TYPE

Consider a signal $f_1(t)$ in the acoustic SLM1 is imaged on a signal $f_2(t)$ in the acoustic SLM2, as shown in Fig. 8.13 [15]. The optical wave passing though SLM1 is given by

$$1 + i\alpha f_1(x - vt)\cos[k_a(x - vt)]. \tag{8.54}$$

In the SLM2, the acoustic wave propagates to the opposite direction of the acoustic wave of SLM1. The optical wave passing through SLM2 is given by

$$1 + i\alpha f_2(x + vt)\cos[k_a(x + vt)]. \tag{8.55}$$

The first diffraction order from wave SLM1 is selected and illuminates SLM2. The first diffraction order wave from SLM2 is selected and the final output intensity is given by

$$I(t) = \alpha^2 \left| \int_{-\infty}^{\infty} f_1(x - vt)f_2(x + vt)dx \right|^2 = \alpha^2 \left| \int_{-\infty}^{\infty} f_1(x)f_2(x + 2vt)dx \right|^2. \tag{8.56}$$

The correlation function $\Phi_{12}(t)$ is obtained as the time signal.

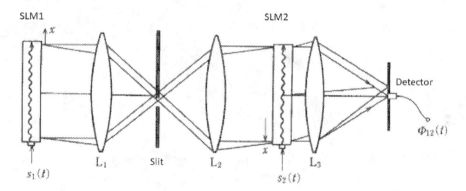

Figure 8.13 Space-integral type correlator.

8.6.2 TIME-INTEGRAL TYPE

The time signal of $s_1(t)$ modulates the intensity of the light source by SLM1 and illuminates SLM2, as shown in Fig. 8.14 [16].

The first order diffraction wave from SLM2 is given by

$$1 + i\alpha f_2(t - x/v)\cos[2\pi v_a(t - x/v)]. \tag{8.57}$$

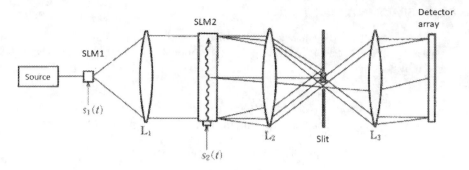

Figure 8.14 Time-integral type correlator.

Integration in time of the zero order diffraction wave with the phase shift of $\pi/2$ and the first order diffraction wave gives

$$I(t) = \int_{-\infty}^{\infty} s_1(t) \left| i + i\frac{\alpha}{2} f_2(t - x/v) \exp[i2\pi v_a(t - x/v)] \right|^2 dt$$

$$= \int_{-\infty}^{\infty} f_1(t) \cos(2\pi v_a t) dt$$

$$+ \alpha \int_{-\infty}^{\infty} f_1(t) f_2(t - x/v) \cos(2\pi v_a t) \cos[2\pi v_a(t - x/v)] dt$$

$$+ \frac{\alpha^2}{4} \int_{-\infty}^{\infty} f_1(t) f_2^2(t - x/v) \cos(2\pi v_a t) dt$$

$$\propto \cos(k_a x) \int_{-\infty}^{\infty} f_1(t) f_2(t - x/v) dt \tag{8.58}$$

The envelope of the carrier $\cos(2k_a x)$ gives the correlation function $\Phi_{12}(x)$, which is detected by a linear photo-sensor array.

8.7 JOINT TRANSFORM CORRELATOR

An alternative method for optical correlators is the joint Fourier transform correlator [17, 18]. Consider two input patterns $f_1(x,y)$ and $f_2(x,y)$, located in the input plane P_1 with a distance d, illuminated by a coherent light, as shown in Fig. 8.15(a). The intensity of Fourier transform of the inputs in the P_2 plane is recorded, for example, with a photographic film. The recorded intensity distribution is given by

$$\begin{aligned}
I(v_x, v_y) &= |\mathscr{F}[f_1(x + d/2, y) + f_2(x - d/2, y)]|^2 \\
&= |F_1(v_x, v_y) \exp(i\pi d v_x) + F_2(v_x, v_y) \exp(-i\pi d v_x)|^2 \\
&= |F_(1(v_x, v_y)|^2 + |F_(2(v_x, v_y)|^2 \\
&\quad + F_1(v_x, v_y) F_2^*(v_x, v_y) \exp(i2\pi d v_x) \\
&\quad + F_1^*(v_x, v_y) F_2(v_x, v_y) \exp(-i2\pi d v_x), \tag{8.59}
\end{aligned}$$

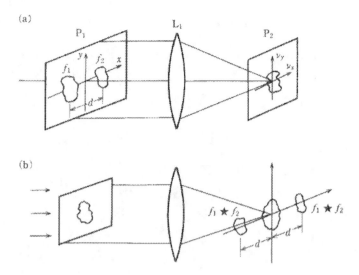

Figure 8.15 Joint transform correlator.

where the Fourier transform of $f_1(x,y)$ and $f_2(x,y)$ are $F_1(v_x,v_y)$ and $F_2(v_x,v_y)$, respectively. The transmittance of the recorded film is given by

$$t(v_x,v_y) = t_0 + \gamma[|F_1|^2 + |F_2|^2]$$
$$+ \gamma[F_1 F_2^* \exp(i2\pi dv_x) + F_1^* F_2 \exp(-i2\pi dv_x)]. \qquad (8.60)$$

By Fourier transforming this by a coherent optical system, we have

$$\mathscr{F}[t(v_x,v_y)] = t_0 \delta(x,y)$$
$$+ \gamma[f_1 \star f_2(x,y) + f_2 \star f_2(x,y)]$$
$$+ \gamma[f_1 \star f_2(x+d,y) + f_1 \star f_2(-x+d,-y)]. \qquad (8.61)$$

This means that two correlation patterns $f_1 \star f_2$ located with a distance d around the optical axis are obtained, as shown in Fig.8.15(b).

8.8   OPTICAL ADDITION AND OPTICAL SUBTRACTION

The optical addition operation of pattern intensities can be easily performed by the detection of two patterns with the same detector or double exposure recording two patterns on a photographic medium. To make amplitude addition, two amplitudes of patterns are recorded as a double exposure hologram and the two images are reconstructed at once so that the superposition of two amplitudes are obtained.

Since the intensity is non-negative, optical subtraction of intensity pattern is not easy. The optical subtraction operations are performed by amplitude subtraction after the intensity distribution of patterns are transformed to the amplitude distributions or the use of nonlinear characteristics of spatial light modulators or photo-detectors [19].

As an example of the former method, the double exposure holographic method is introduced [20].

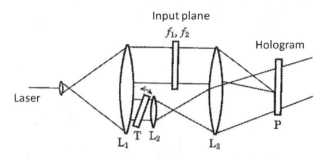

Figure 8.16 Optical subtraction by a hologram.

As shown in Fig. 8.16, the first and second Fourier transform holograms of the pattern $f_1(x,y)$ and $f_2(x,y)$ located in the input plane are recorded as a double exposure hologram, where the reference phase of the second hologram is shifted by an amount of π. The transmittance of the double exposure hologram is given by

$$\begin{aligned}
t(x,y) =& t_0 + \gamma[|F_1(v_x,v_y) + R\exp(i2\pi v_x \sin(\theta/\lambda)|^2 \\
&+ |F_2(v_x,v_y) + R\exp(i2\pi v_x \sin(\theta/\lambda)\exp(-i\pi)|^2] \\
=& t_0 + \gamma[|F_1(v_x,v_y)|^2 + |F_2(v_x,v_y)|^2 + 2|R|^2] \\
&+ \gamma R[F_1(v_x,v_y) - F_2(v_x,v_y)]\exp[-i2\pi v_x \sin(\theta/\lambda)] \\
&+ \gamma R[F_1^*(v_x,v_y) - F_2^*(v_x,v_y)]\exp[i2\pi v_x \sin(\theta/\lambda]
\end{aligned} \tag{8.62}$$

according to Eqs. (7.3) and (7.4). By reconstructing with the first reference wavefront $R\exp(i2\pi v_x \sin\theta/\gamma)$ and Fourier transforming it, the reconstructed image is given by

$$g(x,y) = g_0(x+d,y) + g_1(x,y) + g_{-1}(x+2d,y), \tag{8.63}$$

where $d = \sin\theta/\lambda$,

$$g_0(x,y) = \mathscr{F}\{t_0 + \gamma[|F_1|^2 + |F_2|^2 + 2|R|^2]\} \tag{8.64}$$

$$g_1(x,y) = \gamma R^2[f_1(x,y) - f_2(x,y)] \tag{8.65}$$

$$g_{-1}(x,y) = \gamma R^2[f_1^*(x,y) - f_2^*(x,y)]. \tag{8.66}$$

The intensity of first diffraction component $g_1(x,y)$ gives the difference $|f_1(x,y) - f_2(x,y)|^2$ of the input patterns. Because of the double exposure method, it is not a real time process.

The grating subtraction method, as shown in Fig. 8.17, is a real time method [21]. Consider two input patterns $f_1(x,y)$ and $f_2(x,y)$, located with a distance d. Similar

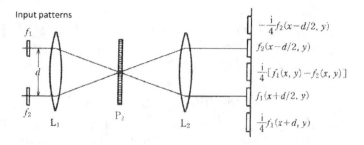

Input patterns

f_1

d

f_2

L_1 P_2 L_2

$-\dfrac{i}{4}f_2(x-d/2,y)$

$f_2(x-d/2,y)$

$\dfrac{i}{4}[f_1(x,y)-f_2(x,y)]$

$f_1(x+d/2,y)$

$\dfrac{i}{4}f_1(x+d,y)$

Figure 8.17 Optical subtraction by a grating.

to the case of the joint transform correlator, the Fourier transform of the two patterns is formed in plane P_2.

$$G(v_x,v_y) = \mathscr{F}\left[f_1\left(x+\frac{d}{2},y\right)+f_2\left(x-\frac{d}{2},y\right)\right]$$
$$=F_1(v_x,v_y)\exp(i\pi dv_x)+F_2(v_x,v_y)\exp(-i\pi dv_x). \qquad (8.67)$$

Then a grating with the pitch Λ in the spatial frequency domain is placed in the plane P_2,

$$H(v_x) = 1 + \frac{1}{2}\cos\left(\frac{2\pi v_x}{\Lambda}+\delta\right), \qquad (8.68)$$

where δ denotes the phase of the grating. By Fourier transforming again with the lens L_2, we have

$$g(x,y) = \mathscr{F}[G(v_x,v_y)\cdot H(v_x)]$$
$$=f_1\left(x+\frac{d}{2},y\right)+f_2\left(x-\frac{d}{2},y\right)$$
$$+\frac{1}{4}\mathscr{F}\left\{F_1(v_x,v_y)\exp\left[-i2\pi\left(\frac{1}{\Lambda}-\frac{d}{2}\right)v_x\right]\cdot\exp(-i\delta)\right.$$
$$+F_2(v_x,v_y)\exp\left[i2\pi\left(\frac{1}{\Lambda}-\frac{d}{2}\right)v_x\right]\cdot\exp(i\delta)$$
$$+F_1(v_x,v_y)\exp\left[i2\pi\left(\frac{1}{\Lambda}+\frac{d}{2}\right)v_x\right]\cdot\exp(i\delta)$$
$$+F_2(v_x,v_y)\exp\left[-i2\pi\left(\frac{1}{\Lambda}+\frac{d}{2}\right)v_x\right]\cdot\exp(-i\delta)\right\}. \qquad (8.69)$$

In the case when $1/\Lambda = d/2$ and $\delta = -\pi/2$, we have

$$g(x,y) = f_1\left(x+\frac{d}{2},y\right)+f_2\left(x-\frac{d}{2},y\right)+\frac{i}{4}[f_1(x,y)-f_2(x,y)]$$
$$-\frac{i}{4}[f_1(x+d,y)-f_2(x-d,y)]. \qquad (8.70)$$

Finally, we have the difference of patterns $|f_1(x,y) - f_2(x,y)|^2$ is obtained around the position $x = 0$ in the output plane. In this method, five patterns near the positions, $x = d, d/2, 0, -d/2, -d$, in the output plane are observed. To increase the separation of the output patterns, the three beam grating method is proposed [22].

8.9   COORDINATE TRANSFORM

In many cases, such as operations of addition, subtraction and correlation between patterns, the normalization of coordinates of patterns is required. The coordinate transform using a phase filter based on a computer-generated hologram has been discussed [23].

Consider the mapping of the point (x, y) in the x-y coordinate system into the point (u, v) in the u-v coordinate system,

$$u = p(x, y) \tag{8.71}$$
$$v = q(x, y). \tag{8.72}$$

Such transforms are obtained as solutions of a certain kind of differential equations. The existence of solutions such as

$$\frac{\partial p(x, y)}{\partial y} = \frac{\partial q(x, y)}{\partial x} \tag{8.73}$$

is required.

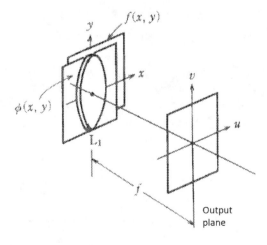

Figure 8.18   Optical coordinate transform system.

Consider an optical setup where the x-y coordinate system and the u-v coordinate system are in Fourier transform each other, as shown in Fig. 8.18 and a filter for coordinate transform is located just after the x-y plane. In the output plane, we have

$$F(u,v) = \iint_{-\infty}^{\infty} f(x,y) \exp[i\phi(x,y)] \exp\left[-i\frac{2\pi}{\lambda f}(xu+yv)\right] dxdy$$

$$= \iint_{-\infty}^{\infty} f(x,y) \exp[ik\Phi(x,y,u,v)] dxdy, \qquad (8.74)$$

where

$$\Phi(x,y,u,v) = \frac{\lambda}{2\pi}\phi(x,y) - \left(\frac{xu}{f} + \frac{yv}{f}\right). \qquad (8.75)$$

It should be noted that the wavenumber $k = 2\pi/\lambda$ is extremely large. Except for the area where $\Phi(x,y)$ is slowly varying, the term $\exp[ik\Phi(x,y)]$ vibrates violently between ± 1, and therefore, this term does not contribute to the integration of Eq. (8.74).[2] The term which contributes the integral is the point

$$\frac{\partial \Phi(x,y)}{\partial x} = \frac{\partial \Phi(x,y)}{\partial y} = 0. \qquad (8.76)$$

Finally, from Eq. (8.75) we have the differential equations of the mapping

$$\frac{\partial \phi}{\partial x} = \frac{2\pi}{\lambda f}u \qquad (8.77)$$

$$\frac{\partial \phi}{\partial y} = \frac{2\pi}{\lambda f}v. \qquad (8.78)$$

8.9.1   EQUAL MAGNIFICATION IMAGING

Since $u = -x, v = -y$ to make the same magnification in imaging, by integrating Eqs. (8.77) and (8.78), we have

$$\phi(x,y) = -\frac{\pi(x^2+y^2)}{\lambda f}. \qquad (8.79)$$

This is the same as the lens described by Eq. (6.13).

---

[2]If the integral of Eq. (8.74) has an extreme value only in the point (x_0,y_0), we have

$$\frac{\partial \Phi}{\partial x} = \Phi_x(x_0,y_0) = 0, \qquad \frac{\partial \Phi}{\partial y} = \Phi_y(x_0,y_0) = 0,$$

$f(x,y)$ is continuous at (x_0,y_0), and

$$\Phi_{xx}\Phi_{yy} - \Phi_{xy}^2 = 0, \qquad \Phi_{xy} \neq 0,$$

where $\Phi_{xx}, \Phi_{xy}, \Phi_{yy}$ are the partial differentials at $(x_0.y_0)$. In this case, if $k \to \infty$

$$\iint_{-\infty}^{\infty} f(x,y) \exp[ik\Phi(x,y,u,v)] dxdy = \frac{i2\pi f(x_0,y_0)}{k\sqrt{\Phi_{xx}\Phi_{yy} - \Phi_{xy}^2}} \exp[ik\Phi(x_0,y_0)]$$

is valid. This method for calculating the integral is called the stationary phase method.

8.9.2   LOGARITHMIC COORDINATE TRANSFORM

In the case of logarithmic coordinate transform such as $u = \ln x, v = \ln y$, we have

$$\phi(x,y) = \frac{2\pi(x\ln x - x + y\ln y - y)}{\lambda f}. \tag{8.80}$$

Figure 8.19 shows a CGH filter of the logarithmic coordinate transform described Eq. (8.80). The coordinate transform using this filter and the optical Fourier transform gives the Mellin transform mentioned in Sec. 8.10.

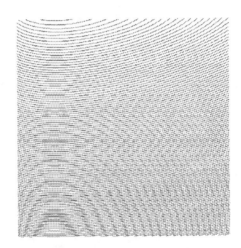

Figure 8.19   CGH for coordinate transform.

8.10   MELLIN TRANSFORM

The Mellin transform is defined by

$$\mathcal{M}[f(x)] = M(s) = \int_0^\infty f(x)x^{s-1}dx \tag{8.81}$$

for the input signal $f(x)$. In this transform, if the magnification of the input signal is changed, the absolute value of its Mellin transform is not changed. The Mellin transform is useful for the scale invariant signal processing, that is, the normalization of the input patterns with different magnifications for spatial frequency filtering [24]. Suppose the Mellin transform is the mapping into the complex space, with a complex number

$$s = \alpha + i\omega \tag{8.82}$$

and $x = \exp(\xi)$, Eq. (8.81) is rewritten as

$$M(s) = \int_{-\infty}^\infty f[\exp(\xi)]\exp[\xi(\alpha + i\omega)]d\xi. \tag{8.83}$$

This is the Laplace transform. Suppose that s is a pure imaginary number, such as $s = -i2\pi v$, Eq. (8.83) is rewritten as

$$M(v) = \int_{-\infty}^{\infty} f[\exp(\xi)] \exp(-i2\pi v\xi) d\xi. \tag{8.84}$$

This is the Fourier transform.

The Mellin transform is also defined as

$$M(v) = \int_{0}^{\infty} f(x)^{-i2\pi v-1} dx \tag{8.85}$$

with a pure imaginary number $s = -i2\pi v$. The inverse Mellin transform exists as

$$f(x) = \mathcal{M}^{-1}[f(x)] = \int_{-\infty}^{\infty} M(v)x^{i2\pi v} dv. \tag{8.86}$$

Next, consider the Mellin transform of $f(\alpha x)$,

$$\mathcal{M}[f(\alpha x)] = \int_{0}^{\infty} f(\alpha x)x^{-i2\pi v-1} dx = \int_{0}^{\infty} f(x)\left(\frac{x}{\alpha}\right)^{-i2\pi v-1} \frac{dx}{\alpha}$$
$$= \alpha^{i2\pi v} \int_{0}^{\infty} f(x)x^{-i2\pi v-1} dx = \alpha^{i2\pi v} \mathcal{M}[f(x)]. \tag{8.87}$$

Finally we have

$$|\mathcal{M}[f(\alpha x)]| = |\mathcal{M}[f(x)]|. \tag{8.88}$$

This means that the absolute value of the Mellin transform is scale invariant.

The 2-D Mellin transform is defined as

$$F(v_x, v_y) = \iint_{-\infty}^{\infty} f(x,y)x^{-i2\pi v_x-1} y^{-i2\pi v_y-1} dxdy. \tag{8.89}$$

Using

$$x = \exp(\xi) \tag{8.90}$$
$$y = \exp(\eta) \tag{8.91}$$

we have

$$F(v_x, v_y) = \iint_{-\infty}^{\infty} f[\exp(\xi), \exp(\eta)] \exp[-i2\pi(v_x\xi + v_y\eta)] d\xi d\eta. \tag{8.92}$$

Therefore, the Fourier transform of $f[\exp(\xi), \exp(\eta)]$ gives the Mellin transform. Since

$$\xi = \ln x \tag{8.93}$$
$$\eta = \ln y, \tag{8.94}$$

the filter for the coordinate transform is given by

$$\phi(x,y) = \frac{2\pi(x\ln x - x + \ln y - y)}{\lambda f} \tag{8.95}$$

using Eq. (8.80). This filter is already shown in Fig. 8.19. The Mellin transform is obtained by coordinate transform of an input pattern using the filter shown in Fig. 8.19 and then optical Fourier transforming. An example of the optical Mellin transform of rectangular apertures with different sizes is shown in Fig. 8.20 [25]. Because of the intensity of the Mellin transform, their sizes are identical.

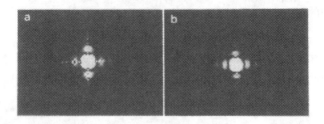

Figure 8.20   Mellin transforms of different size rectangles.

8.11   WAVELET TRANSFORM

The Fourier transform is used to detect the periodicity in signals because the integral kernel $\exp(i2\pi v x)$ is a periodic function. Furthermore, the Fourier transform is used to find the similarity in data because the Fourier spectrum is represented with power series for the data with self-similarity.

It is difficult to discuss local information in Fourier spectrum because the Fourier transform kernel $\exp(i2\pi v x)$ expands in the full range of x. For this reason, to obtain a local spectrum, an input function is limited only in a local area with a window function and Fourier transformed. However, this method reduces the resolution of periodicity because of the localized integral kernel.

The wavelet transform does not preserve the periodicity but preserves the similarity rigorously. This transform is suitable in analysis of local similarity in data because of its local integral kernel and similarity [26].

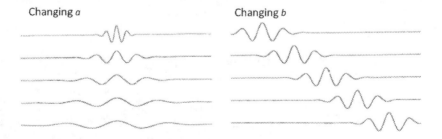

Figure 8.21   Wavelets with a and b changed. The mother wavelet is the Meyer wavelet $m(x) = \exp(i2\pi v_0 x) \cdot \exp(-x^2/2)$, its real part is shown.

At first, consider a function $m(x)$, called a mother wavelet. By using the mother wavelet, the function series with two parameters a and b are defined by

$$h_{a,b}(x) = \frac{1}{\sqrt{a}} m\left(\frac{x-b}{a}\right), \tag{8.96}$$

which is called a wavelet. As shown in Fig. 8.21, wavelets are similar to each other. As compared with the Fourier transform, the parameter a in the wavelet corresponds to the period, inverse of the frequency, but the position parameter b has no corresponding parameter in the Fourier transform. The wavelet transform is defined using kernels of a series of wavelets. Wavelet transforms with continuous parameters a and b are called continuous wavelet transforms, and wavelet transform with discrete parameters are discrete wavelet ones. Discrete wavelet transforms with orthogonal kernels are called orthogonal wavelet transforms. Some examples of mother wavelets are shown in Fig. 8.22.

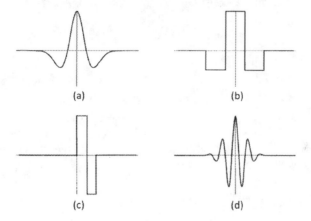

(a)   (b)

(c)   (d)

Figure 8.22 Examples of mother wavelets. (a) Mexican hat wavelet $m(x) = (1 - x^2)\exp(-x^2/2)$, (b) French hat wavelet, (c) Haar wavelet and (d) real part of Meyer wavelet $m(x) = \exp(i2\pi v_0 x) \cdot \exp(-x^2/2)$.

A continuous wavelet transform of a function $f(x)$ is defined by

$$W_f(a,b) = \frac{1}{\sqrt{a}} \int_{-\infty}^{\infty} m\left(\frac{x-b}{a}\right) f(x) dx. \tag{8.97}$$

Since the wavelet transform is a correlation between $m(x)$ and $f(x)$, the wavelet transform can be performed by optical spatial frequency filtering.

The Fourier transform of the wavelet is given by

$$H_{a,b}(v) = \sqrt{a}\exp(-i2\pi bv)M(av), \tag{8.98}$$

where $M(v)$ denotes the Fourier transform of $m(x)$.

Now, consider matched filtering with the wavelet transform. By using Fourier transform of the wavelet transform of the signal $s(x, y)$ to be detected

$$W_s(a_x, a_y, b_x, b_y) = \sqrt{a_x a_y} \iint S(v_x, v_y) \cdot M^*(a_x v_x, a_y v_y)$$
$$\times \exp[\mathrm{i}2\pi(b_x v_x + b_y v_y)]\mathrm{d}v_x \mathrm{d}v_y \qquad (8.99)$$

and the Fourier transform of the wavelet transform of the input signal $f(x, y)$

$$W_f(a_x, a_y, b_x, b_y) = \sqrt{a_x a_y} \iint F(v_x, v_y) \cdot M^*(a_x v_x, a_y v_y)$$
$$\times \exp[\mathrm{i}2\pi(b_x v_x + b_y v_y)]\mathrm{d}v_x \mathrm{d}v_y. \qquad (8.100)$$

The output of the wavelet matched filtering is given by

$$\iint_{-\infty}^{\infty} W_f(a_x, a_y, b'_x, b'_y) W_s^*(a_x, a_y, b'_x - b_x, b'_y - b_y)\mathrm{d}b'_x \mathrm{d}b'_y$$
$$= \iint_{-\infty}^{\infty} S^*(v_x, v_y) \cdot M(a_x v_x, a_y v_y) \cdot F(v_x, v_y) \cdot M^*(a_x v_x, a_y v_y)$$
$$\times \exp[\mathrm{i}2\pi(b_x v_x + b_y v_y)]\mathrm{d}v_x \mathrm{d}v_y. \qquad (8.101)$$

This means that the wavelet matched filter is given by

$$G(v_x, v_y) = S^*(v_x, v_y) \cdot |M(a_x v_x, a_y v_y)|^2. \qquad (8.102)$$

It should be noted that the wavelet matched filter is the conventional matched filter $S^*(v_x, v_y)$ weighted by $|M(a_x v_x, a_y v_y)|^2$ [27].

8.12   X-RAY COMPUTER TOMOGRAPHY

X-ray computer tomography is a technique to obtain a sectional image of specific areas of scanned objects [28]. As shown in Fig. 8.23, many projected images are detected by X-ray projection for an object with different angles. The cross-sectional image of the object is calculated by using many projected images detected. Suppose the absorption coefficient of the object for X-ray used is μ, which is dependent of a kind of the object material, its density, the wavelength of X-ray, and so on. The transmitted X-ray intensity after the object is

$$I = I_0 \exp(-\mu l), \qquad (8.103)$$

where I_0 denotes the input intensity and l the path length of X-ray. The absorption coefficient is given by

$$\mu = -\frac{\log(I/I_0)}{l}. \qquad (8.104)$$

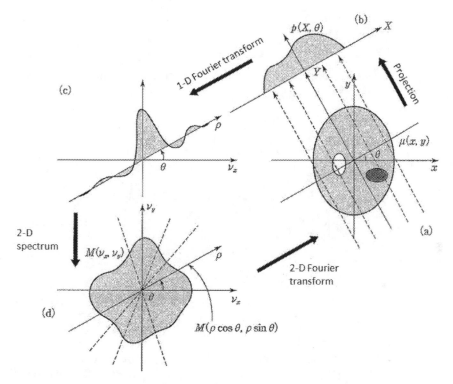

Figure 8.23 Principle of X-ray computer tomography (2-D Fourier transform method). (a) object, (b) projected image, (c) Fourier transform of projected image, and (d) 2-D Fourier spectrum.

Next, consider the absorption distribution $\mu(x,y)$ of the object. The coordinate system (x,y) fixed to the object is defined. The projection direction is defined along the Y axis, and its orthogonal direction is defined as the X axis. The angle θ is defined by the angle between the x axis and the X axis.

$$x = X \cos \theta - Y \sin \theta \tag{8.105}$$
$$y = X \sin \theta + Y \cos \theta. \tag{8.106}$$

The absorption coefficient distribution projected to the direction of the Y axis is given by

$$p(X,\theta) = \int_{-\infty}^{\infty} \mu(x,y)\mathrm{d}Y$$
$$= \int_{-\infty}^{\infty} \mu(X \cos \theta - Y \sin \theta, X \sin \theta + Y \cos \theta)\mathrm{d}Y. \tag{8.107}$$

This is called the Radon transform. Suppose the 2-D Fourier transform of the absorption coefficient $\mu(x,y)$

$$M(\nu_x, \nu_y) = \iint_{-\infty}^{\infty} \mu(x,y) \exp[-i2\pi(x\nu_x + y\nu_y)]dxdy \qquad (8.108)$$

and its polar coordinates $(\rho\cos\theta, \rho\sin\theta)$ version

$$
\begin{aligned}
M(\rho\cos\theta, \rho\sin\theta) &= \iint_{-\infty}^{\infty} \mu(x,y) \exp[-i2\pi\rho(x\cos\theta + y\sin\theta)]dxdy \\
&= \int_{-\infty}^{\infty} \left[\int_{-\infty}^{\infty} \mu(\cos\theta - Y\sin\theta, X\sin\theta + Y\cos\theta)dY \right] \\
&\quad \times \exp(-i2\pi X\rho)dX \\
&= \int_{-\infty}^{\infty} p(X,\theta) \exp(-i2\pi X\rho)dX.
\end{aligned}
\qquad (8.109)
$$

8.12.1   TWO-DIMENSIONAL FOURIER TRANSFORM METHOD

One-D Fourier transform of the projected image $p(X,\theta)$ gives the spectrum component of $M(\rho\cos\theta, \rho\sin\theta)$, which is the ρ axis spectrum component of 2-D Fourier transform $M(\nu_x, \nu_y)$ of the absorption coefficient $\mu(x,y)$. This is called the projection-slice theorem. By using many 1-D projected images with different angles of θ and their 1-D Fourier transforms, all the components of the 2-D Fourier transform of $M(\nu_x, \mu_y)$ are obtained and then its inverse 2-D Fourier transform gives the absorption coefficient of $\mu(x,y)$.

$$\mu(x,y) = \iint_{-\infty}^{\infty} M(\nu_x, \nu_y) \exp[i2\pi(x\nu_x + y\nu_y)]d\nu_x d\nu_y. \qquad (8.110)$$

This method of X-ray tomography is called 2-D Fourier transform method.

8.12.2   FILTERED BACK PROJECTION METHOD

Because 2-D Fourier transform $M(\nu_x, \nu_y)$ is represented in the orthogonal coordinates (ν_x, ν_y) and the obtained spectrum $M(\rho\cos\theta, \rho\sin\theta)$ is given in the polar coordinates (ρ, θ), the coordinate system transform from the polar coordinates to the orthogonal coordinates by interpolation is necessary. Since the angle of θ is discrete, data points in the higher frequency in the orthogonal coordinates system are more sparse. Even though the data points in the higher frequency are interpolated, errors in the higher frequency are serious in many cases. To reduce the interpolation error,

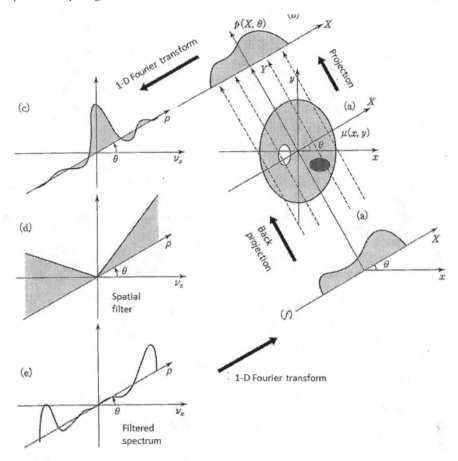

Figure 8.24 Principle of X-ray computer tomography (filtered back projection method). (a) object, (b) projected image, (c) its 1-D Fourier spectrum, (d) spatial filter, (e) filtered spectrum, and (f) its 1-D inverse Fourier transform.

the filtered back projection method, as shown in Fig. 8.24, is introduced. At first, Eq. (8.110) is rewritten with the polar coordinate $(v_x = \rho \cos \theta, v_y = \rho \sin \theta)$.

$$
\begin{aligned}
\mu(x,y) &= \iint_{-\infty}^{\infty} M(v_x, v_y) \exp[i2\pi(xv_x + yv_y)]dv_x dv_y \\
&= \int_0^{2\pi} \int_0^{\infty} M(\rho \cos \theta, \rho \sin \theta) \exp[i2\pi\rho(x\cos \theta + y\sin \theta)]\rho d\rho d\theta \\
&= \int_0^{\pi} \int_{-\infty}^{\infty} M(\rho \cos \theta, \rho \sin \theta)|\rho| \exp[i2\pi\rho(x\cos \theta + y\sin \theta)]d\rho d\theta \\
&= \int_0^{\pi} \left[\int_{-\infty}^{\infty} M(\rho \cos \theta, \rho \sin \theta)|\rho| \exp(i2\pi X\rho)d\rho \right] d\theta.
\end{aligned} \tag{8.111}
$$

The filtered projection image is obtained by the 1-D Fourier transform of the projected image $M(\rho \cos\theta, \rho \sin\theta)$ filtered by a high pass filter of $|\rho|$. This is called the filtered back projection image. This procedure is repeated for many projection angles θ to obtain many filtered projection images. Finally, integrating the filtered projection images by θ gives the absorption coefficient $\mu(x,y)$. The two methods, the two-dimensional Fourier transform method and the filtered back projection method, are identical mathematically.

PROBLEMS

1. Design the optical frequency filter, which converts a pattern $f(x,y)$ into a pattern $g(x,y)$. This is called the code transform filter.
2. The Mellin transform is a scale invariant transform. What is a rotation invariant transform? What is a scale and rotation invariant filter?

BIBLIOGRAPHY

Stark, H., ed. 1982. *Applications of Optical Fourier Transforms*. Academic Press.
Feitelson, D. G. 1988. *Optical Computing*. MIT Press, Cambridge
Arrathoon, R., ed. 1989. *Optical Computing*. Marcel Dekker, New York.

1. Cutrona, L. J., Leith, E. N., Palermo, C. J. and Porcello, I. J. 1960. Optical data processing and filtering systems. *IRE Trans Inf Theory*. IT-6: 386.
2. Vander Lugt, A. 1974. Coherent optical processing. *Proc. IEEE*. 62: 1300.
3. Lee, S. H. 1974. Mathematical Operations By Optical Processing. *Opt. Eng.* 13: 196.
4. Zernike, F. 1935. Phase contrast *Z. Tech. Phys.* 16: 454.
5. Jacquinot, P. and Roizen-Dossier, B. 1964. Apodisation. *Progress in Optics*. Wolf, E. ed. III: 29. North-Holland.
6. Osterberg, H. and Wilkins, Jr., J. E. 1959. The Resolving power of a coated objective. *J. Opt. Soc. Amer.* 39: 553.
7. Vander Lugt, A. 1964. Signal detection by complex spatial filtering. *IEEE Trans Inf Theory*. IT-10: 139.
8. Turin, G. L. 1964. An introduction to matched filters. *IRE Trans Inf Theory*. IT-6: 311.
9. Franks, L. E. 1966. *Signal Theory*. Prentice-Hall.
10. Yatagai, T. 1976. Optimum spatial filter for image restoration degraded by multiplicative noise. *Opt. Commun.* 19: 236.
11. Helstrom, C. W. 1967. Image restoration by the method of least squares. *J. Opt. Soc. Amer.* 57: 297.
12. Casasen, D. P. 1981. *Optical Information Processing*. S. H. Lee ed. Springer-Verlag, Berlin, p. 181.
13. Turpin, T. M. 1981. Spectrum analysis using optical processing. *Proc. IEEE*. 69: 79.
14. Rhodes, W. T. 1981. Acousto-optic signal processing: Convolution and correlation. *Proc. IEEE*. 69: 65.
15. Sprague, R. A. 1977. A review of acousto-optic signal correlators. *Opt. Eng.* 16: 476.
16. Sprague, R. A. and Koliopolus, C. L. 1976. Time integrating acoustooptic correlator. *Appl. Opt.* 15: 89.
17. Weaver, C. and Goodman, J. W. 1960. A Technique for Optically Convolving Two Functions. *Appl. Opt.* 5: 124.

18. Javidi, B. 1989. Nonlinear joint power spectrum based optical correlation. *Appl. Opt.* 28: 2358.
19. Ebersole, J. F. 1975. Optical Image Subtraction. *Opt. Eng.* 14: 436.
20. Gabor, D., Stroke, G. W., Restrick, R., Funkhouser, A. and Brumm, B. 1965. Optical image synthesis (complex amplitude addition and subtraction) by hollographic Fourier transformation. *Phys. Lett.* 18: 116.
21. Lee, S. H., Yao, S. K. and Milnes, A. M. 1970. Optical image synthesis (Complex amplitude addition and subtraction) in real time by a diffraction-grating interferometric method. *J. Opt. Amer.* 60: 1037.
22. Matsuda,K., Takeya,K., Tsujiuchi, J. and Shinoda, M. 1971. An experiment of image-subtraction using holographic beam splitter. *Opt. Commun.* 2: 425.
23. Bryngdahl, O. 1974. Geometrical transformations in optics. *J. Opt. Soc. Amer.* 64: 1092.
24. Casasent, D. and Szczulkowski, C. 1976. Optical Mellin transforms using computer generated holograms. *Opt. Commumn.* 19: 217.
25. Yatagai, T., Choji, K. and Saito, H. 1981. Pattern classification using optical Mellin transform and circular photodiode array. *Opt. Commun.* 38: 162.
26. Chui, C, K. 1992. *An Introduction to Wavelets.* Academic Press.
27. Roberge, D. and Sheng, Y. 1994. Optical wavelet matched filter. *Appl. Opt.* 33: 5285.
28. Saleh, B. 2011. *Introduction to Subsurface Imaging.* Cambridge University Press, Chapter 4.

9 Analytic Signal and Hilbert Transform

The use of the complex representation of a real monochromatic sinusoidal wave $\cos(2\pi\nu t)$ has been mentioned in the previous chapters, and its usability is emphasized. In linear operations, the wave in a complex form is used in the calculation process and the real part of the final result of the operation gives the physically existing wave. In the case of non-monochromatic waves, what kinds of representation of wave is convenient? The analytic signal is introduced to discuss non-chromatic waves. The analytic signal is a complex signal, of which imaginary part is given by the Hilbert transform of its real part.

9.1 COMPLEX REPRESENTATION AND NEGATIVE FREQUENCY

Consider a real sinusoidal signal

$$u(t) = A(t)\cos\psi(t). \tag{9.1}$$

If we can regard it as a sinusoidal wave, its phase is rewritten as

$$\psi(t) = 2\pi\nu t + \phi(t), \tag{9.2}$$

where $A(t)$ and $\phi(t)$ are more slowly-varying functions as compared with $2\pi\nu t$. Differentiating the both sides of Eq. (9.2), we have

$$\frac{d\psi(t)}{dt} = 2\pi\nu + \frac{d\phi(t)}{dt}. \tag{9.3}$$

Since $2\pi\nu \gg d\phi(t)/dt$, the second term can be neglected, we have

$$\nu = \frac{1}{2\pi}\frac{d\psi(t)}{dt} \tag{9.4}$$

The differential of the phase gives the frequency.

Equation (9.1) shows that the signal $u(t)$ sinusoidally vibrates according to the increase of the phase $\psi(t)$. This is considered as the rotation of a vector as shown in Fig. 9.1(a). The projection of the vector to the x-axis gives $A(t)\cos\psi(t)$.

It should be noted that the amplitude $A(t)$ and the phase $\psi(t)$ of the signal $u(t)$ are not estimated independently by using Eq. (9.1). Figure 9.1(a) shows that the length and the direction of the vector are not estimated with its projection to the x-axis, $A(t)\cos\psi(t)$.

To estimate the length and the direction of a vector, its projection to the y-axis, $A(t)\sin\psi(t)$ is necessary. Consider the coordinate system shown in Fig. 9.1(a) to be

DOI: 10.1201/9781003121916-9

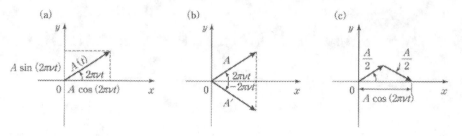

Figure 9.1 Complex representation of sinusoidal signal.

in a complex space, that is, the x-axis is a real axis and the y-axis an imaginary axis. The real signal $u(t)$ is extended to a complex signal $u_c(t)$,

$$u_c(t) = A(t)\cos \psi(t) + iA(t)\sin \psi(t) = A(t)\exp[i\psi(t)], \qquad (9.5)$$

which includes the information on both the length and the direction of the vector. The reason why the wave function is represented by a complex value is that the general complex signal gives the amplitude and phase components more easily than the real signal only on the real axis.

Because of the relation $\cos[\psi(t)] = \cos[-\psi(t)]$, the x-axis projection $A(t)\cos \psi(t)$ is obtained possibly by two vectors, rotated inversely each other, as shown in Fig. 9.1(b). As shown in Fig. 9.1(c), the x-axis projection $A(t)\cos \psi(t)$ is represented by the sum of two vectors. Its complex representation is given by

$$A(t)\cos \psi(t) = \frac{1}{2}\{A(t)\exp[i\psi(t)] + A(t)\exp[-i\psi(t)]\}. \qquad (9.6)$$

The first term in the right side is the vector rotating anticlockwise and the second term the vector rotating clockwise. When the vector rotates with a constant velocity, $\psi(t) = 2\pi vt$, we get

$$u(t) = A(t)\cos(2\pi vt) = \frac{1}{2}[A(t)\exp(i2\pi vt) + A(t)\exp(-i2\pi vt)]. \qquad (9.7)$$

This means that the wave $A(t)\cos(2\pi vt)$ is described by the sum of vectors $A(t)\exp(i2\pi vt)$ and $A(t)\exp(-i2\pi vt)$, rotating inversely to each other. That is, the sum of vectors rotating inversely to each other gives a vector on the real axis, whose value is real.

The rotation direction of a vector is inverted means that its rotation angle is negative, and therefore, its frequency is negative. This is the physical meaning of negative frequency. Equation (9.7) shows that a real sinusoidal function consists of positive and negative frequency components.

9.2 ANALYTIC SIGNAL

The amplitude of a real monochromatic light wave is described by

$$u(t) = a(t)\cos(\phi - 2\pi vt). \tag{9.8}$$

Its complex representation is given by

$$u_c(t) = a(t)\exp[i(\phi - 2\pi vt)]. \tag{9.9}$$

If necessary, its real part gives the real amplitude $u(t)$. Even for non-monochromatic light, it is convenient to use the complex representation so that the real part $\mathrm{Re}[u_c(t)]$ of the complex representation of $u_c(t)$ can give a real amplitude of light wave. This is the analytic signal.

Consider the Fourier integral representation of a signal $v(t)$,

$$v(t) = \int_{-\infty}^{\infty} V(v)\exp(i2\pi vt)dv \tag{9.10}$$

where

$$V(v) = A(v)\exp[i\Phi(v)]. \tag{9.11}$$

$A(v)$ and $\Phi(v)$ denote the amplitude and phase of $V(v)$, respectively. If $v(t)$ is real, we have

$$A(v) = A(-v) \tag{9.12}$$

$$\Phi(-v) = -\Phi(v), \tag{9.13}$$

which are given by the complex conjugate of Fourier transform of Eq. (9.10).

In this case, we have

$$v(t) = 2\int_0^{\infty} A(v)\cos[\Phi(v) + 2\pi vt]dv. \tag{9.14}$$

corresponding to a real signal $v(t)$. Let us introduce a complex function

$$z(t) = 2\int_0^{\infty} A(v)\exp[i\Phi(v) + i2\pi vt]dv. \tag{9.15}$$

Because

$$z(t) = 2\int_0^{\infty} A(v)\cos[\Phi(v) + 2\pi vt]dv$$
$$+ 2i\int_0^{\infty} A(v)\sin[\Phi(v) + 2\pi vt]dv, \tag{9.16}$$

we have

$$v(t) = \mathrm{Re}[z(t)] \tag{9.17}$$

It should be noted that a complex signal $z(t)$ can be described only with positive frequency component. The complex signal $z(t)$ is called the analytic signal of a real signal $v(t)$. The analytic signal $z(t)$ is rewritten as

$$z(t) = z_R(t) + iz_I(t) = v(t) + i\hat{v}(t), \tag{9.18}$$

where $z_R(t)$ and $z_I(t)$ denote the real and imaginary parts of $z(t)$, respectively.

The Fourier transform of the analytic signal $z(t)$ is given by

$$Z(v) = \int_{-\infty}^{\infty} z(t) \exp(-i2\pi vt)dt \qquad (9.19)$$

By using Eq. (9.15), we have

$$\begin{aligned} z(t) &= 2\int_{0}^{\infty} V(v)\exp(i2\pi vt)dv \\ &= 2\int_{-\infty}^{\infty} V(v)U(v)\exp(i2\pi vt)dv. \end{aligned} \qquad (9.20)$$

where

$$U(v) = \begin{cases} 1: & v \geq 0 \\ 0: & v < 0. \end{cases} \qquad (9.21)$$

From Eqs. (9.20) and (9.21), we have the relationship

$$Z(v) = \begin{cases} 2V(v): & v \geq 0 \\ 0: & v < 0. \end{cases} \qquad (9.22)$$

Here we rewrite Eq. (9.22)

$$Z(v) = V(v) + i\hat{V}(v), \qquad (9.23)$$

where

$$\begin{aligned} \hat{V}(v) &= \begin{cases} -iV(v): & v \geq 0 \\ iV(v): & v < 0 \end{cases} \\ &= -iV(v)\mathrm{sgn}(v). \end{aligned} \qquad (9.24)$$

Functions $\hat{V}(v)$ and $\hat{v}(t)$ are Fourier transform pairs. The relationship of $Z(v)$ with $V(v)$ and $i\hat{V}(v)$ is shown in Fig. 9.2.

The analytic signal is useful in complex representation of non-chromatic optical waves, as described in Sec. 10.1.

9.3 HILBERT TRANSFORM

From Eqs. (9.18) and (9.24), we have

$$z_I(t) = \hat{v}(t) = \mathscr{F}^{-1}[\hat{V}(v)] \qquad (9.25)$$

$$= \mathscr{F}^{-1}\{[V(v)[-i\,\mathrm{sgn}(v)]\} \qquad (9.26)$$

$$= z_R(t) * \frac{1}{\pi t} \qquad (9.27)$$

$$= \frac{1}{\pi} P \int_{-\infty}^{\infty} \frac{z_R(t')}{t-t'}dt', \qquad (9.28)$$

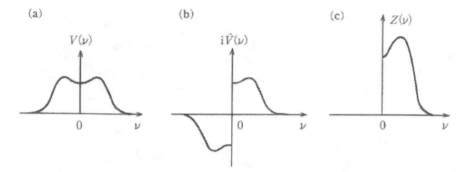

Figure 9.2 Fourier spectrum $V(v)$ and analytic signal $Z(v)$ of $z(t)$, and also $i\hat{V}(v)$.

where P means the Cauchy principal value.[1] This means that the real part $z_R(t) = v(t)$
and imaginary part $z_I(t) = \hat{v}(t)$ are not independent, each other. The transform from
$z_R(t)$ to $z_I(t)$ is called the Hilbert transform.

$$\hat{v}(t) = \mathcal{H}[v(t)] = v(t) * \frac{1}{\pi t}, \tag{9.29}$$

where $\mathcal{H}[\cdot]$ denotes the Hilbert transform.

Table 9.1 shows examples of Hilbert transforms. The Hilbert transform has its inverse transform. The Hilbert transform of $\cos(2\pi vt)$ is $\sin(2\pi vt)$ and then its Hilbert
transform is $-\cos(2\pi vt)$, as shown in Fig. 9.3.

Table 9.1
Functions and Their Hilbert Transforms

Function $v(t)$	Hilbert transform $\hat{v}(t) = \mathcal{H}[v(t)]$
$\hat{v}(t)$	$-v(t)$
$v(t)\exp(i2\pi vt)$	$-iv(t)\exp(i2\pi vt)$
$v(t)\exp(-i2\pi vt)$	$iv(t)\exp(-i2\pi vt)$
$v(t)\cos(2\pi vt)$	$v(t)\sin(2\pi vt)$
$v(t)\sin(2\pi vt)$	$-v(t)\cos(2\pi vt)$
$\delta(t)$	$1/(\pi t)$

---

[1] The Cauchy principal value is defined as

$$\int_{-\infty}^{\infty} \frac{f(x)}{x-y}dx = \lim_{\varepsilon \to 0}\left[\int_{-\infty}^{y-\varepsilon} \frac{f(x)}{x-y}dx + \int_{y-\varepsilon}^{\infty} \frac{f(x)}{x-y}dx\right]$$

and is the integral value without divergence at $x = y$.

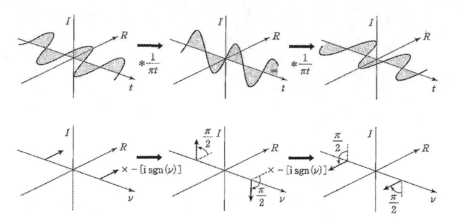

Figure 9.3   Hilbert transform of $\cos(2\pi t)$ is obtained by the convolution of $\cos(2\pi t)$ and $1/(\pi t)$. Its further Hilbert transform gives $-\cos(2\pi t)$. The upper part shows this process and the lower one its Fourier transform.

Consider a Hilbert transform pair, $v(t)$ and $\hat{v}(t)$.

$$
\begin{aligned}
\int_{-\infty}^{\infty} v(t) \cdot \hat{v}^*(t)\,\mathrm{d}t &= \int_{-\infty}^{\infty} \mathscr{F}^{-1}[V(v) * \hat{V}^*(-v)]\,\mathrm{d}t \\
&= \iiint_{-\infty}^{\infty} V(v-v')\hat{V}^*(-v')\,\mathrm{d}v'\exp(\mathrm{i}2\pi vt)\,\mathrm{d}v\,\mathrm{d}t \\
&= \iint_{-\infty}^{\infty} v(t)\exp(\mathrm{i}2\pi v't)\,\mathrm{d}t\,\hat{V}^*(-v')\,\mathrm{d}v' \\
&= \int_{-\infty}^{\infty} V(-v')\hat{V}^*(-v')\,\mathrm{d}v' \\
&= \int_{-\infty}^{\infty} (-\mathrm{i})\mathrm{sgn}(v')V(-v')V^*(-v')\,\mathrm{d}v' \\
&= \mathrm{i}\int_{-\infty}^{\infty} \mathrm{sgn}(v)|V(v)|^2\,\mathrm{d}v \\
&= 0
\end{aligned}
\tag{9.30}
$$

This means that the real and imaginary parts are orthogonal to each other. The real part of an analytic signal is called its in-phase component and the imaginary part is the quadrature part, in some cases. Since

$$
A(t)\exp(\mathrm{i}2\pi vt) = A(t)\cos(2\pi vt) + \mathrm{i}A(t)\sin(2\pi vt) \tag{9.31}
$$

and the Hilbert transform of $A(t)\cos(2\pi vt)$ is $A(t)\sin(2\pi vt)$, $A(t)\exp(\mathrm{i}2\pi vt)$ is the analytic signal of $A(t)\cos(2\pi vt)$.

   If a signal $A(t)$ with a limited range is modulated in the frequency v, this modulated signal is written as $A(t)\cos(2\pi vt)$, whose analytic signal $z(t)$, its in-phase component $z_R(t)$ and quadrature component $z_I(t)$ are shown in Fig. 9.4.

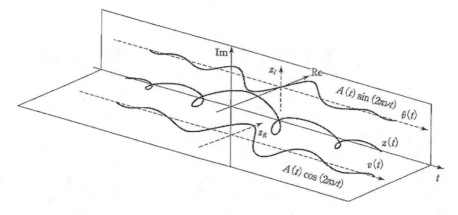

Figure 9.4 Analytic signal and its in-phase and quadrature component.

According to the properties of the analytic signal, we have directly the amplitude $A(t)$ and phase $\phi(t)$ of the modulated signal $A(t)\cos[2\pi vt + \phi(t)]$ by the real and imaginary parts of its analytic signal $z(t)$.

$$A(t) = \sqrt{z_R(t)^2 + z_I(t)^2} \tag{9.32}$$

$$\phi(t) = \tan^{-1}\frac{z_I(t)}{z_R(t)} \tag{9.33}$$

An application of the Hilbert transform to interference fringe analysis is described in Sec. 10.5.

PROBLEMS

1. Determine the Hilbert transform of $\hat{v}(t)$, where $\hat{v}(t)$ is the Hilbert transform of a continuous time signal $v(t)$.
2. Determine the Hilbert transform of $m(t)\cos(2\pi v_0 t)$.
3. For the Hilbert transform pair, $v(t)$ and $\hat{v}(t)$, prove that their power spectrum (square of spectrum) are equal.
4. From an amplitude modulated signal $v(t) = m(t)\cos(2\pi v_0 t)$, we can obtain the signal $m(t)$ by using the Hilbert transform. Discuss how to detect the original signal $m(t)$.

BIBLIOGRAPHY

Papoulis, A. 1968. *Systems and Transforms with Applications in Optics*. McGraw-Hill, New York.

Goodman, J. W. 1985. *Statistical Optics*. John Wiley & Sons, New York, p. 104.

Bracewell, R. N. 1986. *The Fourier Transform and Its Applications*. 2nd ed. McGraw-Hill, New York.

10 Coherence, Spectroscopy and Fringe Analysis

The major aspects of light and its propagation have been mentioned previously. The ideal cases of point and monochromatic light sources have also been discussed. In this chapter, some topics on light sources, such as optical coherence related to the size of light sources and their spectral properties are discussed. A consideration of light source is now applied to measure the angular diameter of stars, while its spectroscopic measurement method is known as the Fourier transform spectroscopy. The phase shift interferometry will bel also discussed in relation with the Fourier transform spectroscopy and some fringe analysis techniques. Finally, the use of the Hilbert transform in the interferometric fringe analysis is presented.

10.1 COHERENCE

Consider an extended and non-monochromatic light source, as shown in Fig. 10.1. The coherence between the arbitrary points P_1 and P_2 in the space is estimated by the interference effect observed at a point P' behind the plane of P_1 and P_2 [1, 2]. To estimate the interference effect, the interference fringes between the wavefront from two pinholes at P_1 and P_2 are observed in the plane of P'. Because of a non-monochromatic wave, the wavefront $v(r,t)$ is described by using an analytic signal, which is described in Sec. 9.2. The amplitude of the wave at P' is given by

$$v(r_{P'}, t_1, t_2) = v(r_1, t_1) + v(r_2, t_2), \tag{10.1}$$

where $v(r_1, t_1)$ and $v(r_2, t_2)$ denote the complex amplitudes of the waves at P_1 and P_2, respectively.

In the case of a stationary light source, $v(r', t_1, t_2)$ is not a function of independent variables of t_1 and t_2 but its difference $\tau = t_2 - t_1$ corresponds the time delay between the arriving time of the waves emitted from P_1 and P_2 at the same time. Therefore Eq. (10.1) is rewritten as

$$v(r_{P'}, t_1, t_2) = v(r_1, t + \tau) + v(r_2, t). \tag{10.2}$$

The intensity at P' is given by

$$I(r') = \langle v(r', t) v^*(r', t) \rangle, \tag{10.3}$$

where $\langle \cdots \rangle$ denotes the operation of time averaging. By submitting Eqs. (10.2) into (10.3),

$$I(r') = I(r_1) + I(r_2) + 2\text{Re}[\Gamma_{12}(r_1, r_2, \tau)], \tag{10.4}$$

DOI: 10.1201/9781003121916-10

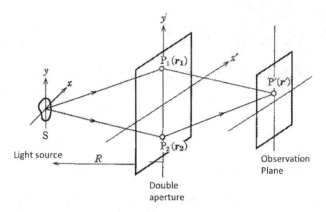

Figure 10.1   Definition of coherence.

where

$$\Gamma_{12}(r_1, r_2, \tau) = \langle v(r_1, t+\tau) v^*(r_2, t) \rangle, \tag{10.5}$$

which is rewritten as

$$\Gamma_{12}(r_1, r_2, \tau) = \lim_{T \to \infty} \frac{1}{2T} \int_{-T}^{T} v(r_1, t+\tau) v^*(r_2, t) \mathrm{d}t. \tag{10.6}$$

It should be noted that Γ_{12} is the complex cross-correlation between the waves at P_1 and P_2, which is called the mutual coherence function in physical optics.

In the case of $r_1 = r_2$,

$$\Gamma_{11}(r_1, r_1, \tau) = \langle v(r_1, t+\tau) v^*(r_1, t) \rangle \tag{10.7}$$

is called the auto coherence function. Further, if $\tau = 0$,

$$\Gamma_{11}(0) = \langle v(r_1, t) v^*(r_1, t) \rangle = I(r_1), \tag{10.8}$$

which gives the intensity at the point r_1.

The normalized mutual coherence function Γ_{12} by using Γ_{11} and Γ_{22} is given by

$$\gamma_{12}(\tau) = \frac{\Gamma_{12}(\tau)}{\sqrt{\Gamma_{11}(0)}\sqrt{\Gamma_{22}(0)}} = \frac{\Gamma_{12}(\tau)}{\sqrt{I(r_1)}\sqrt{I(r_2)}}. \tag{10.9}$$

The intensity given by Eq. (10.4) is rewritten as

$$I(r') = I(r_1) + I(r_2) + 2\sqrt{I(r_1)I(r_2)}\mathrm{Re}[\gamma_{12}]. \tag{10.10}$$

By using the Schwarz inequality,

$$0 \leq |\gamma_{12}(\tau)| \leq 1 \tag{10.11}$$

is proved. As mentioned in Sec. 10.1, the light is coherent under the condition $\gamma = 1$, otherwise incoherent $\gamma = 0$. The light with $0 < \gamma < 1$ is partially coherent.

Since $\gamma_{12}(\tau)$ is complex,

$$\gamma_{12}(\tau) = |\gamma_{12}(\tau)| \exp[i\phi_{12}(\tau)]. \tag{10.12}$$

Equation (10.10) is rewritten as

$$I(r') = I(r_1) + I(r_2) + 2\sqrt{I(r_1)I(r_2)}|\gamma_{12}|\cos[\phi_{12}(\tau)]. \tag{10.13}$$

By using the definition of the visibility Eq. (2.13)

$$V = \frac{I_{max} - I_{min}}{I_{max} + I_{min}}, \tag{10.14}$$

and Eq. (10.13), the visibility is given by

$$V = \frac{2\sqrt{I(r_1)I(r_2)}}{I(r_1) + I(r_2)}|\gamma_{12}(\tau)|. \tag{10.15}$$

This gives the accurate form of Eq. (2.15).

10.1.1 TEMPORAL COHERENCE

In the case when the observation point P_1 coincides with P_2, the mutual coherence function $\Gamma_{12}(r_1, r_2, \tau)$ becomes the auto-coherence function $\Gamma_{12}(r, \tau)$. Because of the correlation of a wave at a point in different times and a function depending on only the time difference τ, this is called temporal coherence. The integral representation of Eq. (10.7) is rewritten as

$$\Gamma_{11}(\tau) = \int_{-\infty}^{\infty} v(t+\tau)v^*(t)dt \tag{10.16}$$

and its Fourier transform is

$$\mathscr{F}[\Gamma_{11}(\tau)] = \int_{-\infty}^{\infty} \Gamma_{11} \exp(-i2\pi v\tau)d\tau. \tag{10.17}$$

Then by using Eq. (10.16), we have

$$\begin{aligned}
\mathscr{F}[\Gamma_{11}(\tau)] &= \iint_{-\infty}^{\infty} v(t+\tau)v^*(t)dt \cdot \exp(-i2\pi v\tau)d\tau \\
&= \int_{-\infty}^{\infty} v^*(t)\exp(-i2\pi vt)dt \cdot \int_{-\infty}^{\infty} v(t+\tau)\exp[-i2\pi v(t+\tau)]d\tau \\
&= |F(v)|^2
\end{aligned} \tag{10.18}$$

where the spectrum of the wave is defined as

$$F(v) = \int_{-\infty}^{\infty} v(t)\exp(-i2\pi vt)dt. \tag{10.19}$$

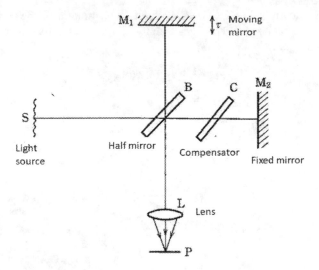

Figure 10.2 Spectrum interferometer used for temporal coherence analysis.

The Fourier transform of the autocorrelation $\Gamma_{11}(\tau)$ is equal to the power spectrum $|F(v)|^2$. This is the Wiener-Khintchine theorem.

In order to measure the temporal coherence, a Michelson type interferometer, as shown in Fig. 10.2, is used. Changing the optical path length enables to change the time difference τ. The visibility V is measured as a function of the time difference τ, and then $\Gamma_{11}(\tau)$ is evaluated. Its Fourier transform gives the power spectrum of the light source. This is the principle of Michelson's spectrum interferometry [3].

The coherence time Δt is defined as the time difference where the visibility decreases to zero. The rigorous definition of the coherence time Δt is

$$(\Delta t)^2 = \frac{\int_{-\infty}^{\infty} \tau^2 |\Gamma_{12}(\tau)|^2 d\tau}{\int_{-\infty}^{\infty} |\Gamma_{12}(\tau)|^2 d\tau} \tag{10.20}$$

and the spectrum width Δv, by using the power spectrum $G(v)$,

$$(\Delta v)^2 = \frac{\int_0^{\infty} (v - \bar{v})^2 G(v)^2 d\tau}{\int_0^{\infty} G(v)^2 d\tau}, \tag{10.21}$$

where the central frequency $\bar{v}$ is defined as

$$\bar{v} = \frac{\int_0^{\infty} \bar{v} G(v) d\tau}{\int_0^{\infty} G(v) d\tau}. \tag{10.22}$$

The following relationship is derived

$$\Delta t \Delta v \geq \frac{1}{2}. \tag{10.23}$$

This is the same as the uncertainty principle in quantum mechanics.

10.1.2 SPATIAL COHERENCE

The spatial coherence is defined by the correlation $\gamma_{12}(0)$ at two different points, P_1 and P_2, in the same time ($\tau = 0$). Consider the spatial coherence of an extended light source, as shown in Fig. 10.1. Suppose the light source is quasi-monochromatic and different points in the extended light source are independent, that is, mutually incoherent. The quasi-monochromatic wave is represented

$$v(t) = a(t)\exp(-i2\pi\hat{v}t),\tag{10.24}$$

where $\hat{v}$ denotes the mean frequency. The extended source is divided into small areas as small as a point source. The m-th area is denoted by $d\sigma_m$. Let light waves from the small area $d\sigma_m$ to points P_1 and P_2 are denoted by $v_{m1}(t)$ and $v_{m2}(t)$. The mutual correlation function is given by

$$\Gamma_{12}(\tau) = \left\langle \left[\sum_{m1} v_{m1}(t+\tau)\right]\left[\sum_{m2} v_{m2}^*(t)\right]\right\rangle.\tag{10.25}$$

Because of no contributions from $m1 \neq m2$, we have

$$\Gamma_{12}(\tau) = \sum_m \langle v_{m1}(t+\tau)v_{m2}^*(t)\rangle.\tag{10.26}$$

Since the light waves from the area $d\sigma_m$ to points P_1 and P_2 are given by

$$v_{mj}(t) = a\left(t - \frac{r_{mj}}{c}\right)\frac{\exp[-i2\pi\hat{v}(t-r_{mj}/c)]}{r_{mj}} \quad (j=1,2).\tag{10.27}$$

Their time average is given by

$$\langle v_{m1}(t+\tau)v_{m2}^*(t)\rangle$$
$$= \left\langle a(t+\tau-r_{m1}/c)a^*(t-r_{m2}/c)\right\rangle\exp\left[\frac{i2\pi\hat{v}(r_{m1}-r_{m2})}{c}\right]\Big/r_{m1}r_{m2}.\tag{10.28}$$

Since points inside the extended source in $\tau = 0$ are incoherent with each other, we have

$$\left\langle a(t-r_{m1}/c)a^*(t-r_{m2}/c)\right\rangle = I(\sigma)d\sigma_m,\tag{10.29}$$

where $I(\sigma)$ denotes the intensity per unit area in the light source. Finally we have

$$\Gamma_{12}(0) = \int_S I(\sigma)\frac{\exp[i2\pi(r_1-r_2)/\bar{\lambda}]}{r_1 r_2}d\sigma,\tag{10.30}$$

where $\bar{\lambda} = c/\bar{v}$ denotes the average wavelength.

The degree of mutual coherence is written by

$$\gamma_{12}(0) = \frac{1}{\sqrt{\Gamma_{11}(0)\Gamma_{22}(0)}}\int_S I(\sigma)\frac{\exp[i2\pi(r_1-r_2)/\bar{\lambda}]}{r_1 r_2}d\sigma.\tag{10.31}$$

Next, consider the case when the distance between P_1 and P_2 and the size of the extended light source are small enough as compared with the distance from the light source to the plane of the double aperture. In such a case, the diffraction is approximated to be the Fraunhofer diffraction and then we have

$$\gamma_{12}(0) = \int_{-\infty}^{\infty} I(x,y) \exp[i2\pi(xv_x + yv_y)]dxdy, \tag{10.32}$$

where the coordinates in the plane of the light source are (x,y), the positions P_1 and P_2 are (x_1', y_1') and (x_2', y_2'), $r_1 = r_2 = R$, and

$$v_x = \frac{x_1' - x_2'}{\bar{\lambda}R}, \qquad v_y = \frac{y_1' - y_2'}{\bar{\lambda}R} \tag{10.33}$$

The degree of the mutual coherence is equal to the Fourier transform of the intensity distribution of the light source. This is known as the van Cittert-Zernike theorem. The degree of spatial coherence gives the size of the light source. The Michelson stellar interferometer shown in Fig. 10.3 is based on this theorem [3]. Michelson supposed that the star was a disc with homogeneous intensity distribution. By changing the spacing L of the double aperture, the spatial coherence $\gamma_{12}(0)$ was measured, and then the angular diameter α of the star was estimated. The degree of coherence is given by

$$\gamma_{12} = \frac{2J_1\left(\frac{\pi\alpha}{\bar{\lambda}}L\right)}{\frac{\pi\alpha}{\bar{\lambda}}L}. \tag{10.34}$$

where $J_1(\dots)$ is the first-order Bessel function of the first kind. By increasing L, the first point of $\gamma_{12} = 0$ was obtained at L_0, and then

$$\frac{\pi\alpha}{\bar{\lambda}}L_0 = 3.83. \tag{10.35}$$

Finally, we have

$$\alpha = \frac{1.22\bar{\lambda}}{L_0}. \tag{10.36}$$

Using this, Michelson estimated the angular diameter of the star Betelgeuse of Orion to be 0.047 arc second.

10.2 FOURIER TRANSFORM SPECTROSCOPY

The combination of the Michelson spectrum interferometer and the Wiener-Khinchin theorem gives the power spectrum of the light source. Due to the progress of data processing technology, an interferometric spectroscopy has become practical. This method is called the Fourier transform spectroscopy. Figure 10.4 shows an example of its optical setup. The interference intensity is given by

$$I_\sigma(h) = 2B(\sigma)[1 + \cos(2\pi\sigma h)], \tag{10.37}$$

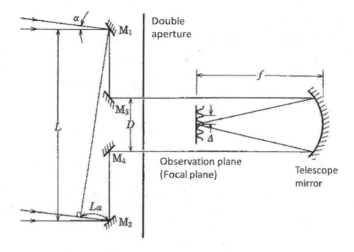

Figure 10.3 Michelson's stellar interferometer.

where the intensity of the light source is $B(\sigma)$ for the wavenumber σ,[1] and the optical path difference is h.

For the light source with the wide spectrum distribution, the intensity is written as $B(\sigma)d\sigma$ in the narrow wavenumber range $\sigma + d\sigma$. The interference intensity is given by

$$dI_\sigma(h) = 2B(\sigma)[1 + \cos(2\pi\sigma h)]d\sigma, \qquad (10.38)$$

where $B(\sigma)$ is the intensity of the wave with the wavenumber σ and is called the spectrum intensity. For all wavenumbers

$$I(h) = 2\int_0^\infty B(\sigma)[1 + \cos(2\pi\sigma h)]d\sigma$$

$$= 2\int_0^\infty B(\sigma)d\sigma + 2\int_0^\infty B(\sigma)\cos(2\pi\sigma h)d\sigma. \qquad (10.39)$$

The polychromatic interference fringe intensity $I(h)$ as a function of the optical path difference h is called an interferogram. In the case of $h = 0$, we have

$$I(0) = 4\int_0^\infty B(\sigma)d\sigma \qquad (10.40)$$

and therefore, we have

$$I(h) - \frac{I(0)}{2} = 2\int_0^\infty B(\sigma)\cos(2\pi\sigma h)d\sigma. \qquad (10.41)$$

Then we have

$$B(\sigma) = 2\int_0^\infty \left[I(h) - \frac{I(0)}{2}\right]\cos(2\pi\sigma h)dh. \qquad (10.42)$$

---

[1] The wavenumber here, is called the spectroscopic wavenumber, is defined by $\sigma = 1/\lambda$.

The Fourier transform of the interferogram without the bias term gives the intensity of the spectrum $B(\sigma)$. Figure 10.5 shows the interferogram of polystyrene and its spectrum.

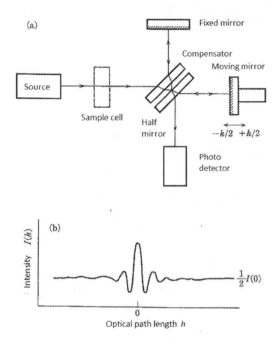

Figure 10.4 Fourier transform spectroscopy. (a) Optical system and (b) interferogram.

The Fourier transform spectrometer is more efficient in the use of input light than the other spectroscopic devices, like prisms and gratings. The efficiency is measured by the etendue, which is defined by the product of the area of the entrance aperture and the solid angle the source subtends as seen from the aperture. In the far infrared region, where it is difficult to use high-power light sources and highly efficient light detectors, the Fourier transform spectroscopy is an indispensable tool.

Let one of the mirrors be tilted at a small angle α. In this case, the intensity of the wave with the wavenumber σ is given by

$$I_\sigma(x) = 2B(\sigma)[1 + \cos(2\pi\sigma\alpha x)] \tag{10.43}$$

as similar to Eq. (10.37). In the case of polychromatic light source, we have

$$I(x) = 2\int_0^\infty B(\sigma)[1 + \cos(2\pi\sigma\alpha x)]\mathrm{d}\sigma. \tag{10.44}$$

$I(x)$ is a function of the real coordinate x, which can be detected by a photodetector array. Fourier transforming of $I(x)$ gives the spectrum $B(\sigma)$, as same as Eq. (10.42) [4]. The method does not have high etendue, but time-varying spectrum can be measured because of no mechanical moving components.

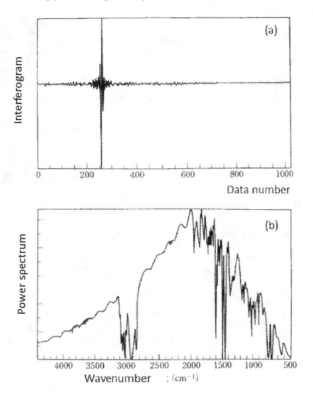

Figure 10.5   Interferogram of (a) polystyrene (b) and its spectrum.

10.3   PHASE SHIFT IN INTERFEROMETRY

Interferometers have been used as a tool to measure surface shape of optical devices and precision mechanical components. A typical interferometer is the Fizeau interferometer, as shown in Fig. 10.6(a). A collimated beam illuminates a reference beam splitter, of which reflected beam is for a reference wave and the other used for illumination of the object surface. An example of a fringe pattern for a silicon wafer is shown in Fig. 10.6(b). The wavelength $\lambda = 0.63$ μm and the spacing of fringe is $\lambda/2 = 0.32 \lambda$ m.

Next, let change the optical path difference in the interferometer, like the Fourier transform interferometer. The method of phase measurement from the change of interferometric fringes is called the phase shift interferometry or the fringe scan method [5, 6, 7]. For simplicity, consider the Twyman-Green interferometer as shown in Fig. 10.7. With the optical path $h(x,y)$ of the test surface due to reflection[2] and the shifted optical path l, the intensity of the interferometric fringe is given by

$$I(x,y) = a(x,y) + b(x,y)\cos\left\{\frac{2\pi}{\lambda}[h(x,y) - l]\right\},   \qquad (10.45)$$

---

[2]$h(x,y)$ is equal to two times of the surface shape.

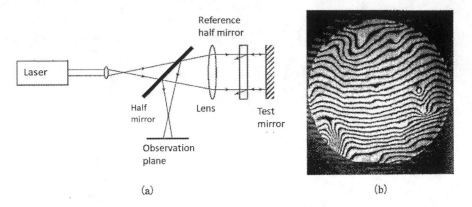

(a)   (b)

Figure 10.6   (a) Fizeau interferometer (b) and its fringe pattern measured.

where $a(x,y)$ and $b(x,y)$ denote the bias term of the fringe intensity and its contrast, respectively.

To obtain the phase $2\pi h(x,y)/\lambda$ of the surface from known information of the intensity $I(x,y,l)$, the terms $a(x,y)$ and $b(x,y)$ should be known or should be eliminated.

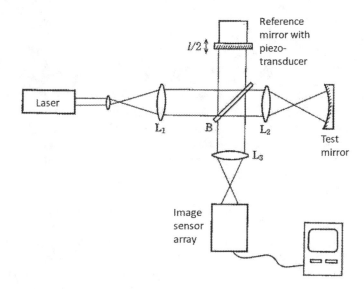

Figure 10.7   Phase shift interferometer.

From Eq. (10.45),

$$I(x,y,l) = a(x,y)$$
$$+ b(x,y)\cos\left\{\frac{2\pi}{\lambda}[h(x,y)]\right\}\cos\left(\frac{2\pi}{\lambda}l\right)$$
$$+ b(x,y)\sin\left\{\frac{2\pi}{\lambda}[h(x,y)]\right\}\sin\left(\frac{2\pi}{\lambda}l\right). \tag{10.46}$$

By Fourier transforming Eq. (10.46),

$$\hat{I}(x,y,v) = \mathscr{F}[I(x,y,l)]$$
$$= a(x,y)\int_{-\infty}^{\infty}\exp(i2\pi lv)dl$$
$$+ b(x,y)\cos\left\{\frac{2\pi}{\lambda}[h(x,y)]\right\}\int_{-\infty}^{\infty}\cos\left(\frac{2\pi l}{\lambda}\right)\exp(i2\pi lv)dl$$
$$+ b(x,y)\sin\left\{\frac{2\pi}{\lambda}[h(x,y)]\right\}\int_{-\infty}^{\infty}\sin\left(\frac{2\pi l}{\lambda}\right)\exp(i2\pi lv)dl$$
$$= a(x,y)\delta(v)$$
$$+ b(x,y)\cos\left\{\frac{2\pi}{\lambda}[h(x,y)]\right\}\frac{[\delta(v-\frac{1}{\lambda})+\delta(v+\frac{1}{\lambda})]}{2}$$
$$+ ib(x,y)\sin\left\{\frac{2\pi}{\lambda}[h(x,y)]\right\}\frac{[\delta(v-\frac{1}{\lambda})-\delta(v+\frac{1}{\lambda})]}{2}. \tag{10.47}$$

By using only the term of $v = \frac{1}{\lambda}$, which corresponds to the first peak in the Fourier spectrum, its real and imaginary parts are

$$\mathrm{Re}[\hat{I}(x,y,v)] = \frac{1}{2}b(x,y)\cos\left\{\frac{2\pi}{\lambda}[h(x,y)]\right\} \tag{10.48}$$

and

$$\mathrm{Im}[\hat{I}(x,y,v)] = \frac{1}{2}b(x,y)\sin\left\{\frac{2\pi}{\lambda}[h(x,y)]\right\}, \tag{10.49}$$

respectively. Finally we have

$$h(x,y) = \frac{\lambda}{2\pi}\tan^{-1}\frac{\mathrm{Im}[\hat{I}(x,y,v)]}{\mathrm{Re}[\hat{I}(x,y,v)]}. \tag{10.50}$$

On the other hand, from Eq. (10.47), the real and imaginary parts of Fourier transform of the interferometric fringe are given in a different form by

$$\int_{-\infty}^{\infty}I(x,y,l)\cos\left(\frac{2\pi l}{\lambda}\right)dl = \frac{1}{2}b(x,y)\cos\left\{\frac{2\pi}{\lambda}[h(x,y)]\right\} \tag{10.51}$$

and

$$\int_{-\infty}^{\infty}I(x,y,l)\sin\left(\frac{2\pi l}{\lambda}\right)dl = \frac{1}{2}b(x,y)\sin\left\{\frac{2\pi}{\lambda}[h(x,y)]\right\}, \tag{10.52}$$

respectively.

Thus, we have

$$h(x,y) = \frac{\lambda}{2\pi} \tan^{-1} \frac{\int_{-\infty}^{\infty} I(x,y,l) \sin(2\pi l/\lambda) dl}{\int_{-\infty}^{\infty} I(x,y,l) \cos(2\pi l/\lambda) dl} \tag{10.53}$$

If the optical path l is changed from 0 to λ at N times with an equal value λ/N. Then, the n-th optical path is given by

$$l_n = \frac{n}{N}\lambda, \quad (n = 0, 1, ..., N-1). \tag{10.54}$$

Since $I(x,y,l)$ is a periodic function of λ to l, we have

$$\mathrm{Re}[\hat{I}(x,y,l)] = \frac{2}{N} \sum_{n=0}^{N-1} I(x,y,l_n) \cos\left(\frac{2\pi n}{N}\right) \tag{10.55}$$

and

$$\mathrm{Im}[\hat{I}(x,y,l)] = \frac{2}{N} \sum_{n=0}^{N-1} I(x,y,l_n) \sin\left(\frac{2\pi n}{N}\right). \tag{10.56}$$

By submitting Eqs. (10.55) and (10.56) into Eq. (10.50),

$$h(x,y) = \frac{\lambda}{2\pi} \tan^{-1} \frac{\sum_{n=0}^{N-1} I(x,y,l_n) \sin(2\pi n/N)}{\sum_{n=0}^{N-1} I(x,y,l_n) \cos(2\pi n/N)}. \tag{10.57}$$

In the most important case of $N = 4$,

$$h(x,y) = \frac{\lambda}{2\pi} \tan^{-1} \frac{I_3 - I_1}{I_0 - I_2}. \tag{10.58}$$

Because Eqs. (10.57) and (10.58) do not include the bias term $a(x,y)$ and the contrast

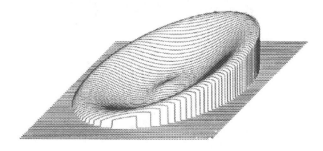

Figure 10.8   Example of measurement of an aspheric mirror by phase shift interferometry.

term $b(x,y)$, this algorithm is not affected by the effects of these terms. An example of an aspheric surface measured is shown in Fig. 10.8.

10.4   FOURIER TRANSFORM FRINGE ANALYSIS

As similar to the case of Eq. (10.43) in the Fourier transform spectroscopy, tilt in the reference mirror is introduced to the interferometer as shown in Fig. 10.7. The intensity of the interferometric fringe is represented by

$$I(x,y) = a(x,y) + b(x,y)\cos\frac{2\pi}{\lambda}[h(x,y) - \alpha x]. \tag{10.59}$$

The fringe intensity includes the carrier term of period λ/α. By Fourier transforming Eq. (10.59), we have

$$\mathscr{F}[I(x,y)] = A(v_x, v_y) + B\left(v_x - \frac{\alpha}{\lambda}, v_y\right) + B^*\left(v_x + \frac{\alpha}{\lambda}, v_y\right), \tag{10.60}$$

where

$$A(v_x, v_y) = \mathscr{F}[a(x,y)] \tag{10.61}$$

$$B(v_x, v_y) = \mathscr{F}\left\{\frac{1}{2}b(x,y)\exp\left[i\frac{2\pi}{\lambda}h(x,y)\right]\right\}. \tag{10.62}$$

Figure 10.9 shows an example of the fringe with a carrier (a) and its Fourier transform (b). By shifting only the first peak of the spectrum to the zero-order and then Fourier transforming it, we have the term

$$c(x,y) = b(x,y)\exp\left[i\frac{2\pi}{\lambda}h(x,y)\right]. \tag{10.63}$$

Let its real and imaginary parts of $c(x,y)$ be $c_R(x,y)$ and $c_I(x,y)$, respectively. Finally, we have the surface profile

$$h(x,y) = \frac{\lambda}{2\pi}\tan^{-1}\left[\frac{c_I(x,y)}{c_R(x,y)}\right], \tag{10.64}$$

as shown in Fig. 10.9(c). This method is called the Fourier transform method in fringe analysis [8].

Consider the equispaced parallel fringe, such as Young's fringe

$$I(x) = a(x) + b(x)\cos(2\pi\alpha x + \phi), \tag{10.65}$$

where ϕ denotes the initial phase of the fringe. In the former case, the spatial distribution of the fringe pattern is considered, but ϕ is constant in this case. By Fourier transforming Eq. (10.65), we have

$$\mathscr{F}[I(x)] = A(v) + \frac{1}{2}B(v - \alpha)\exp(i\phi) + \frac{1}{2}B(v + \alpha)\exp(-i\phi), \tag{10.66}$$

where $A(v)$ and $B(v)$ denote the Fourier transform of $a(x)$ and $b(x)$, respectively. If $b(x)$ is almost constant or symmetry, $B(v)$ is real, and therefore, the phase of the first peak gives the value of ϕ [9]. This method is used in determination of the phase of the reference wave [10] and the absolute length measurement of the block gauge.

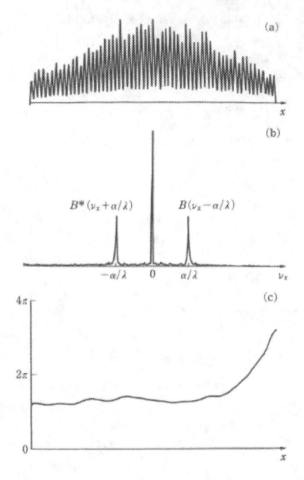

Figure 10.9 (a) Interferometric fringe profile with carrier, (b) its Fourier transform and (c) surface profile analyzed.

10.5   FRINGE ANALYSIS BY HILBERT TRANSFORM

The fringe intensity with a carrier is given by

$$I(x) = a(x) + b(x)\cos[2\pi v_x x + \phi(x)] \tag{10.67}$$

as similar to Eq. (10.59). The method of evaluating the phase $\phi(x)$ by using Fourier transform is presented in Sec. 10.4, and a method using Hilbert transform here. At first the bias term in the fringe pattern is eliminated, by subtracting the average from the fringe pattern or by high-pass filtering of the fringe pattern to delete the lower frequency component. By Hilbert transforming the fluctuating component $v(x) = b(x)\cos[2\pi v_x x + \phi(x)]$, according to Eq. (9.28), the quadrature component $\hat{v}(x) =$

$b(x)\sin[2\pi v_x x + \phi(x)]$ is obtained. Finally we have the phase

$$2\pi v_x x + \phi(x) = \tan^{-1}\left[\frac{\hat{v}(x)}{v(x)}\right]. \tag{10.68}$$

The desired phase $\phi(x)$ is obtained by subtracting the carrier frequency component $2\pi v_x x$. The procedure of the fringe analysis using Hilbert transform is shown in Fig. 10.10.

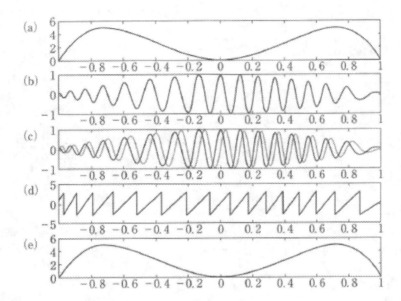

Figure 10.10 Fringe analysis by Hilbert transform. (a) Phase of the fringe ($\phi(x) = -10x^4 + 10x^2$), (b) fluctuating component of fringe profile with carrier ($v(x) = b(x)\cos[2\pi v_x x + \phi(x)]$), (c) in-phase component (solid line) and quadrature component (gray line) given by Hilbert transform, (d) analyzed phase profile and (e) final phase $\phi(x)$ after phase unwrapping and eliminating the carrier $2\pi v_x x$.

PROBLEMS

1. Determine the visibility and the degree of coherence of the light, of which the intensity of the interference fringe is shown in Fig. 10.11.
2. In the Fourier transform spectroscopy, false peaks may be included in the calculated spectrum from N sampling data points of an interferogram. Discuss the reason of false peaks and their countermeasures.
3. In the Fourier transform spectroscopy, discuss the effects in the interferogram of a dispersive object.
4. Discuss the nonlinear effect of the photo-detector in the phase shifting interferometer.

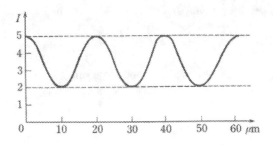

Figure 10.11   Problem 10.1.

5. Estimate the error of the calculated phase in the case the reference phase is not correctly modulated.
6. The heterodyne interferometer is known as an interferometer with very high accuracy. Discuss the difference and similarity between the heterodyne and the phase shifting interferometer in principle.

BIBLIOGRAPHY

Marathay, A. S. 1982. *Elements of Optical Coherence Theory*. John Wiley & Sons, New York.
Goodman, J. W. 1985. *Statistical Optics*. John Wiley & Sons, New York. Chapters 5 and 6.

REFERENCES

1. Born. M. and Wolf, E. 1980. *Principles of Optics*. 6th ed. p. 491, Pergamon Press.
2. Goodman, J. W. 1985. *Statistical Optics*. p. 157, John Wiley & Sons, New York.
3. Michelson, A. A. 1968. *Studies in Optics*. University of Chicago Press.
4. Okamoto, T., Kawata, S. and Minami, S. 1985. Optical method for resolution enhancement in photodiode array Fourier transform spectroscopy. *Appl. Opt.* 24: 4221.
5. Bruning, J. H., Herriott, D. R., Gallagher, J. E., Rosenfeld, D. P., White, A. D. and Brangaccio, D. J. 1974. Digital Wavefront Measuring Interferometer for Testing Optical Surfaces and Lenses, *Appl. Opt.* 13: 2693.
6. Creath, K. 1988. *Progress in Optics*. E. Wolf ed. XXVI: 351. Amsterdam.
7. Schwieder, J. 1990. *Progress in Optics*. E. Wolf ed., XXVII: 273. Amsterdam.
8. Takeda, M. 1982. Fourier-transform method of fringe-pattern analysis for computer-based topography and interferometry. *J. Opt. Soc. Amer.* 72: 156.
9. Lai, G. M. and Yatagai, T. 1994. Use of the fast Fourier transform method for analyzing linear and equispaced Fizeau fringes. *Appl. Opt.* 33: 5935.
10. Lai, G. M. and Yatagai, T. 1991. Generalized phase-shifting interferometry. *J. Opt. Soc. Amer.* 8: 822.

11 Spatio-Temporal Signal Processing

There are many analogies between spatial and temporal optical computing techniques. In this chapter, the use of spatial computing techniques for temporal signal processing is discussed. A very short temporal pulse is Fourier transformed as a spatial image, which is processed by spatial optical processing. Diffraction grating is commonly used in this purpose. Its function as a temporal to spatial signal converter is discussed. The use of the joint Fourier transform correlator in spatio-temporal signal processing using the diffraction grating is presented. Femtosecond light pulse shaper, optical depth measurement and spectral holography are discussed. The spectral domain optical coherence tomography (OCT) is also discussed.

11.1 FEMTOSECOND PULSE SHAPER

11.1.1 FUNCTION OF GRATING

A femtosecond pulse, for example, its pulse width of 10 fs, is not a monochromatic wave because its wavelength range is $\Delta\lambda = 200$ nm. Here an incident wave after a grating is discussed quantitatively [1]. Consider a plane wave of a frequency v_0 with its incident angle γ_0 and its diffraction angle θ_0. The incident angle for the light of a frequency v is tilted by a small angle $\Delta\gamma$ and then $\gamma = \gamma_0 - \Delta\gamma$. Suppose its frequency is not single and its frequency v is shifted for the central frequency v_0, the angular frequency is given by

$$\omega = 2\pi(v - v_0) \tag{11.1}$$

Let us define the angular shift in this case as

$$\Delta\theta = \alpha\Delta\gamma + \beta\omega. \tag{11.2}$$

This means that the diffraction angle depends on the expansion of the incident angle and the bandwidth of the incident wave.

Next two parameters α and β are discussed. Consider a grating as shown in Fig. 11.1. A plane wave comes to the grating groove points A and B with an incident angle γ and is diffracted to the direction of a diffraction angle θ. The optical path difference between two light rays is given by

$$\Delta l = \overline{BC} - \overline{AD} = d\left[\cos\left(\frac{\pi}{2} - \gamma\right) - \cos\left(\frac{\pi}{2} - \theta\right)\right] = d(\sin\gamma - \sin\theta), \tag{11.3}$$

where d denotes a pitch of the grating. Then we have the condition of the m-th order diffraction

$$\sin\gamma - \sin\theta = \frac{m\lambda}{d}. \tag{11.4}$$

DOI: 10.1201/9781003121916-11

Differentiating Eq. (11.4) gives

$$\cos\gamma_0\Delta\gamma - \cos\theta_0\Delta\theta = 0. \tag{11.5}$$

Then α in Eq. (11.2) is

$$\alpha = \frac{\cos\gamma_0}{\cos\theta_0}. \tag{11.6}$$

From Eq. (11.4) for the wavelength change $\Delta\lambda$, we have

$$-\cos\theta_0\Delta\theta = \frac{m}{d}\Delta\lambda \tag{11.7}$$

and because of $\lambda_0\omega_0 = 2\pi c$, we have

$$\Delta\lambda = \frac{2\pi c}{\omega_0^2}\omega, \tag{11.8}$$

because ω means the shift from ω_0. Finally, by using Eqs. (11.7) and (11.8), the coefficient β in Eq. (11.2) is given by

$$\beta = \frac{2\pi cm}{\omega_0^2 d\cos\theta_0}. \tag{11.9}$$

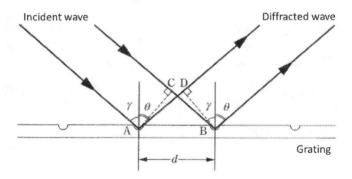

Figure 11.1 Diffraction by grating.

11.1.2 DIFFRACTED BEAM

Suppose that the incident wave propagates to the z-direction, which is the direction of γ_0 to the grating and its amplitude is $a(x,t)$, where the x axis is defined as the direction normal to the propagation direction of the z axis. The Fourier transform of $a(x,t)$ in time and its inverse Fourier transform are given by

$$\tilde{a}(x,\omega) = \int a(x,t)\exp(-i\omega t)\mathrm{d}t \tag{11.10}$$

and

$$a(x,t) = \frac{1}{2\pi} \int \tilde{a}(x,\omega) \exp(i\omega t) d\omega. \tag{11.11}$$

The Fourier transform in space and its inverse Fourier transform are

$$\tilde{A}(\xi,\omega) = \int \tilde{a}(x,\omega) \exp(-i2\pi\xi x) dx \tag{11.12}$$

and

$$\tilde{a}(x,\omega) = \int \tilde{A}(\xi,\omega) \exp(i2\pi\xi x) d\xi. \tag{11.13}$$

This means that the incident wave is a superposition of many plane waves $\tilde{A}(\xi,\omega) \exp(i2\pi\xi x)$ with different propagation directions. The wave propagating in the direction changed by $\Delta\gamma$ to the z axis is given by

$$a(z,x,\omega) = \exp\left\{i\frac{2\pi}{\lambda}[\sin(\Delta\gamma)x + \cos(\Delta\gamma)z]\right.. \tag{11.14}$$

For a small $\Delta\gamma$, for the plane wave $\exp(i2\pi\xi_0 x)$ with a spatial frequency of ξ_0, we have

$$2\pi\xi_0 x = \frac{2\pi}{\lambda} \sin(\Delta\gamma)x = \frac{2\pi}{\lambda}\Delta\gamma x \tag{11.15}$$

by referring Eq. (11.13). Finally, we have

$$\Delta\gamma = \lambda\xi_0. \tag{11.16}$$

This is the relation between the spatial frequency and the direction of the wave propagation (the angle from the z-axis).

As shown in Eq. (11.2), the wave $a_1(x_1,\omega)$ is inclined at an angle of $\alpha\Delta\gamma$ from θ_0. This means that the spatial frequency increases due to Eq. (11.15), where the axis x_1 is defined normal to the propagation direction of the diffracted wave. In order that the diffracted wave propagates with the spatial frequency ξ, the spatial frequency of the corresponding incident wave should be ξ/α. Therefore, the components of the diffracted beam should be

$$\tilde{A}_1(\xi,\omega) = b\tilde{A}\left(\frac{\xi}{\alpha},\omega\right), \tag{11.17}$$

where b is a constant. Inverse Fourier transform of Eq. (11.17) gives the diffraction wave $\tilde{a}_1(x_1,\omega)$, but the component of the diffraction beam includes the effect of the wavelength dispersion $\beta\omega$. Since $\beta\omega$ changes the propagation angle by $\Delta\theta$, the spatial frequency is finally given by

$$\xi' = \frac{\Delta\theta}{\lambda} = \frac{\beta\omega}{\lambda} \tag{11.18}$$

in correspondence with Eq. (11.16). The phase shift due to the wavelength dispersion is

$$2\pi\xi'x_1 = 2\pi\frac{\beta\omega}{\lambda}x_1 = k\beta\omega x_1. \tag{11.19}$$

Finally, we have

$$\tilde{a}_1(x_1,\omega) = \int \tilde{A}_1(\xi,\omega)\exp(\mathrm{i}k\beta\omega x_1)\exp(\mathrm{i}2\pi\xi x_1)\mathrm{d}\xi$$

$$= b\exp(\mathrm{i}k\beta\omega x_1)\int \tilde{A}\left(\frac{\xi}{\alpha},\omega\right)\exp(\mathrm{i}2\pi\xi x_1)\mathrm{d}\xi$$

$$= b'\exp(\mathrm{i}k\beta\omega x_1)\tilde{a}(\alpha x_1). \tag{11.20}$$

11.2 SPATIAL FREQUENCY FILTERING FOR ULTRA-SHORT PULSE

Consider the optical system of a grating lens pair with the femtosecond light incidence, as shown in Fig. 11.2 [2].

The focal lengths of the lens L_1 and L_2 are equally f, and the grating constants of grating G_1 and G_2 are also equally d. The gratings G_1 and G_2 are set symmetrically. The incident angle γ_0 and the diffraction angle θ_0 are also at the same angle to the gratings. The x axis is defined normal to the direction of the incident beam $a_{in}(x,t)$ into the grating G_1, and the x_1 axis is defined normal to the direction diffracted of the beam $a_1(x_1,t)$. Similarly, the input beam $a_3(x_3,t)$ and the diffraction beam $a_{out}(x_4,t)$ are defined. The focal plane of the lens L_1 is the plane H and its coordinate axis is defined as x_2.

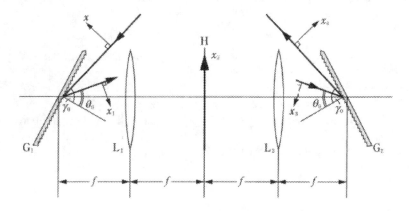

Figure 11.2 Femtosecond pulse shaper by grating-lens pair.

The input beam $a_{in}(x,t)$ is represented separately by the spatial part and temporal part

$$a_{in}(x,t) = a_{in}(x) \cdot a_{in}(t) \tag{11.21}$$

and its temporal spectrum is given by

$$\tilde{a}_{in}(x,\omega) = a_{in}(x) \cdot \tilde{a}_{in}(\omega). \tag{11.22}$$

Since the beam just after diffraction from the grating G_1 is represented by Eq. (11.20), we have

$$\tilde{a}_1(x_1, \omega) = b' \exp(ik\beta\omega x_1)\tilde{a}_{in}(\alpha x_1, \omega). \tag{11.23}$$

After spatially Fourier transformed, the beam at the plane H is given by

$$\tilde{A}_2(\chi, \omega) = b'\tilde{a}_{in}(\omega) \int \tilde{a}_{in}(\alpha x_1) \exp(ik\beta\omega x_1) \exp(-i\chi x_1) dx_1$$

$$= b'\tilde{a}_{in}(\omega) A_{in}\left(\frac{\chi - k\beta\omega}{\alpha}\right), \tag{11.24}$$

where

$$A_{in}(\chi) = \int \tilde{a}_{in}(x_1) \exp(-i\chi x_1) dx_1 \tag{11.25}$$

$$\chi = \frac{2\pi x_2}{\lambda f}. \tag{11.26}$$

A spatial filter $H(\chi)$ is placed in the plane H. Fourier transforming the transmitted wave by the lens L_2, we have the wave on the grating G_2

$$\tilde{a}_3(x_3, \omega) = \frac{1}{2\pi} \int \tilde{A}_2(\chi, \omega) H(\chi) \exp(i\chi x_3) d\chi$$

$$= b''\tilde{a}_{in}(\omega) \int A_{in}\left(\frac{\chi - k\beta\omega}{\alpha}\right) H(\chi) \exp(i\chi x_3) d\chi. \tag{11.27}$$

The wave diffracted by the grating G_2 is given by

$$\tilde{a}_{out}(x_4, \omega) = b''\tilde{a}_3\left(\frac{x_4}{\alpha}, \omega\right) \exp(-ik\beta\omega x_4). \tag{11.28}$$

By using Eq. (11.27), we have

$$\tilde{a}_{out}(x_4, \omega) = b_{out}\tilde{a}_{in}(\omega) \int A_{in}\left(\frac{\chi - k\beta\omega}{\alpha}\right) H(\chi) \exp\left(i\frac{\chi x_4}{\alpha}\right) \exp(-ik\beta\omega x_4) d\chi. \tag{11.29}$$

Finally, we have

$$a_{out}(x_4, t) = b_{out} \frac{1}{2\pi} \int \tilde{a}_{out}(x_4, \omega) \exp(i\omega t) d\omega$$

$$= b'_{out} \iint A_{in}\left(\frac{\chi - k\beta\omega}{\alpha}\right) \tilde{a}_{in}(\omega) H(\chi) \exp\left(i\frac{\chi x_4}{\alpha}\right)$$

$$\times \exp(-ik\beta\omega x_4) \exp(i\omega t) d\omega d\chi. \tag{11.30}$$

Because Eq. (11.30) is the Fourier transform of the product of $A_{in}(\chi)$ and $H(\chi)$ by χ, this corresponds to Eq. (8.2) in spatial frequency filtering. By an appropriately designed filter, the time-frequency filtering to the input pulse can be performed and the pulse shaping and the pulse compression are possible. It should be noted that temporal and spatial components are related complexly, as compared with usual spatial frequency filtering.

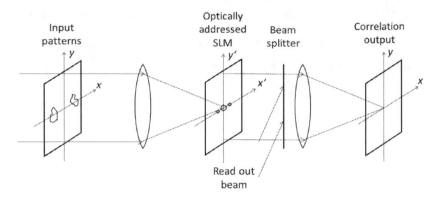

Figure 11.3   Schematic setup of conventional joint transform correlator.

11.3   SPATIO-TEMPORAL JOINT FOURIER TRANSFORM CORRELATOR

In Sec. 8.7 of the joint Fourier transform correlator, two input patterns are located in the input plane, and then two input patterns are jointly Fourier transformed so that the cross correlation between two input patterns are obtained, as shown in Fig. 11.3. The spatial power spectrum of an input image is written on an optically addressed spatial light modulator (SLM), then the spectrum is read out by continuous-wave (cw) light and then spatially Fourier transformed by a lens. Therefore, the correlation of the input image is obtained on the output plane.

On the other hand, in the optical pulse shaper, the input consists of two pulses and this input time signal is Fourier transformed by a grating lens pair, as shown in Fig. 11.4. The temporal power spectrum of input pulses is written on an optically addressed SLM, then the spectrum is read out by the spectrum of output pulses.

Consider a Fourier transform lens of spatial optical computing replaced by a grating lens pair in temporal optical computing. By this replacement, an optical setup for spatial JCT (joint transform correlator) in Fig. 11.3 can be converted to the optical setup in Fig. 11.4.

In the optical setup, the wavelength spectrum of an input pulse is spread on SLM by a grating lens pair. Then the spectrum is converted to its intensity and projected on an optically addressed SLM. The intensity of the spectrum modulates the spectrum of a read-out light pulse on the SLM and the modulated spectrum is transformed to a temporal pulse by another grating lens pair. After the reconstruction, the pulse shape of the reconstructed pulse forms an auto-correlation of the temporal pulse shape of the input pulse.

It is obvious that on the SLM, both the spatial spectrum and wavelength spectrum are spread spatially in the optical setups in Figs. 11.3 and 11.4. Therefore, it will be possible that the right half of Fig. 11.4 is replaced by the right half of Fig. 11.3 [3].

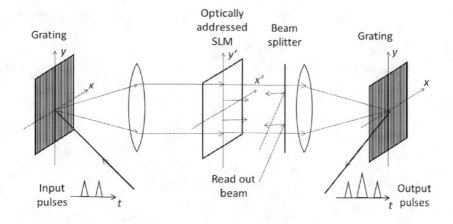

Figure 11.4 Schematic setup of temporal joint transform correlator.

The replaced setup is shown in Fig. 11.5. The optical power spectrum for femtosecond double pulses is recorded on the optically addressed SLM and then this power spectrum is read out by a cw laser light and Fourier transformed spatially. Therefore, the temporal correlation of the input pulses is spread on the output plane spatially. Figure 11.6 shows the experimental result on the autocorrelation measurement of a twin pulse. The horizontal axis is temporal delay, shown as the spatial shift.

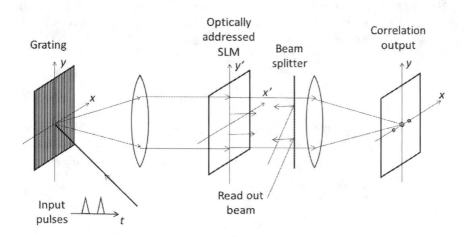

Figure 11.5 Schematic setup of spatio-temporal joint transform correlator.

Similar to Eq. (11.23) in Sec. 11.2, the diffracted wave from the grating is written as

$$\tilde{a}'(x,\omega) = b'\exp(ik\beta\omega x)\tilde{a}(\alpha x,\omega).\tag{11.31}$$

To obtain the temporal expression, the temporal Fourier transform of Eq. (11.31) gives

$$a(x,t) = \frac{1}{2\pi}\int \tilde{a}'(x,\omega)\exp(i\omega t)d\omega$$
$$= b'a(\alpha x, t+k\beta x) = b's(\alpha x)\cdot t(t+k\beta x),\tag{11.32}$$

where $s(x)$ and $t(t)$ denote the spatial and temporal components of $a(\alpha x, t+k\beta x)$, respectively. The spatial Fourier transform of this signal is given by

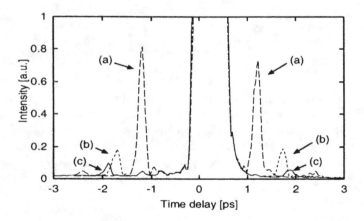

Figure 11.6   The auto-correlation of twin pulses with (a) 1.02-ps, (b) 1.38-ps and (c) 1.53-ps separations, which are measured by the spatio-temporal JTC. We can confirm that the correlation peaks are shifting with the separation of the input pulses.

$$a_H(\chi) \propto S\left(\frac{\chi}{\alpha}\right) * T\left(\frac{\chi}{k\beta}\right),\tag{11.33}$$

where $S(\chi)$ and $T(\chi)$ denote the Fourier spectra of $s(x)$ and $t(t)$, respectively. The size of $s(x)$ is determined by the extension of the diffracted beam from the grating. If the size of $s(x)$ is large enough, $S(\chi)$ can be considered as a delta function. In this case, we have

$$a_H(\chi) \propto T\left(\frac{\chi}{k\beta}\right).\tag{11.34}$$

This means the Fourier transform of the temporal shape of ultra short pulses is obtained as a spatial distribution. The condition that the Fourier transform of the temporal shape of light pulses is obtained as a spatial form is given by

$$\frac{1}{\alpha D} \ll \frac{1}{cT},\tag{11.35}$$

where the diameter of the light beam to the grating and the pulse width of ultra short pulses are denoted by D and T, respectively. When the spectrum width Δv is wide enough, the extent of $S(\chi)$ becomes smaller enough than the extent of $T(\chi)$ and since $\Delta v = 1/T$,

$$\frac{1}{\alpha D} \ll \frac{\Delta v}{c} \tag{11.36}$$

is the necessary condition that the Fourier transform of the temporal shape of light pulses is obtained.

11.4   OPTICAL COHERENCE TOMOGRAPHY

In OCT, a white light interferometer is used to encode the depth information of an object. If the object has a layered structure, the temporal complex amplitude of the reflected wave is modulated by the layered structure. When few micrometer layered information is encoded to the reflected beam, the temporal profile changes very rapidly because a depth of 1 μm is encoded in 6.6 fs, for example, and furthermore when we use a cw broadband light source, the depth information is encoded in the phase of the light, not intensity. This means that the depth profile of an object is not measured directly even if a photodetector has enough high response speed.

To detect the fast signal by a slow detector and to measure the invisible phase changing, a wide bandwidth interferometer is used, as shown in Fig. 11.7. A reflected beam from the reference mirror M_1 and the object beam from a layered structure are superimposed to obtain interference fringes. Since the light source is not monochromatic, the interference fringes are observed only in the case when the optical path difference Δx is zero, as shown in Fig. 11.8. Such localized fringes are called white light fringes. In this optical setup, by shifting the reference mirror position and detecting the peak position of the white light fringes, the layered structure of the object can be measured. This method is called OCT.

In the advanced OCT called Fourier domain optical coherence tomography (FDOCT), the mechanical scanning of the reference mirror is not necessary [4]. Consider that the output of the interfered beam is spectrally resolved by a grating and lens pair and the obtained spatial spectrum is detected by a CCD, as shown in Fig. 11.9.

The detected signal is numerically Fourier transformed to obtain the layered structure of the object. From Eq. (11.34), the spatial distribution of the temporal signal obtained in the focal plane of the lens is $T(\chi/k\beta)$. Since the frequency χ is given by Eq. (11.18), we have

$$\frac{\chi}{k\beta} = \frac{\omega}{2\pi}. \tag{11.37}$$

This means that $T(\chi/k\beta)$ can be considered to be $T(\omega)$. Therefore, the power spectrum detected by CCD is given by

$$I(\omega) = |T_0(\omega) + T(_r(\omega)|^2$$
$$= |T_0(\omega)|^2 + |T_r(\omega)|^2 + T_0(\omega)T_r^*(\omega) + T_0^*(\omega)T_r(\omega), \tag{11.38}$$

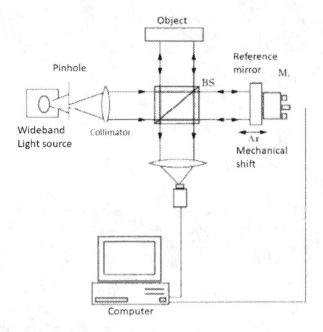

Figure 11.7   Wideband interferometer.

Figure 11.8   White light fringe.

where the spectrum of the object and the spectrum of the reference beam are denoted by $T_0(\omega)$ and $T_r(\omega)$, respectively. Numerical Fourier transform of the power spectrum $I(\omega)$ gives

$$\mathscr{F}[I(\omega)] = T_0(t) * T_0^*(t) + T_r(t) * T_r^*(t) + T_0(t) * T_r^*(t) + T_0^*(t) * T_r(t). \quad (11.39)$$

The third and fourth terms are the correlation between the object and reference beams and its complex conjugate, respectively, and include the depth information or the layered structure Δx of the object. Figure 11.10 summarizes the processing of this method; (a): the intensity of the spectrum detected by CCD, (b): the power spectrum corresponding to Eq. (11.38), (c): the intensity of its Fourier transform, (d): the depth structure corresponding to the third term in Eq. (11.39), (e): the depth structures

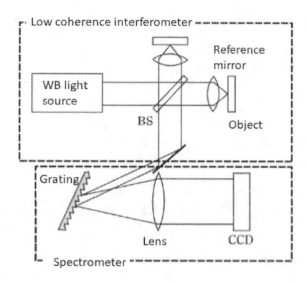

Figure 11.9 Spectral interferometer.

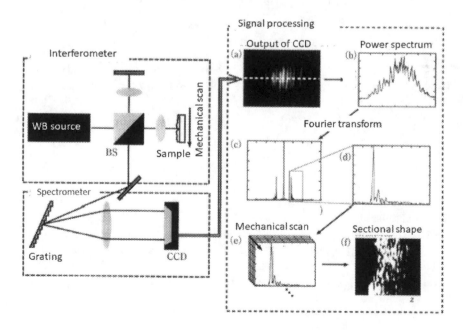

Figure 11.10 Optical system and data processing of Fourier domain OCT.

obtained by mechanical scanning of the sample and (f): the final sectional shape of
the sample, synthesized by the depth structures of (e).

11.5   SPECTRAL HOLOGRAPHY

In spectral interferometry, the spectra of the object and reference beams are superimposed in online and interfered, as shown in Fig. 11.9. The method of the off-axis
superposition of the object and reference spectra is the spectral holography [5]. The
optical system of the spectral holography is shown in Fig. 11.11. Referring to Eq.
(11.38), the intensity detected in the hologram plane is represented by

$$
\begin{aligned}
I(\omega) &= |T_0(\omega)\exp(ik_0r) + T_r(\omega)\exp(ik_rr)|^2 \\
&= |T_0(\omega)|^2 + |T_r(\omega)|^2 + T_0(\omega)T_r^*(\omega)\exp[i(k_0 - k_r)r] \\
&\quad + T_0^*(\omega)T_r(\omega)\exp[-i(k_0 - k_r)r],
\end{aligned}
\tag{11.40}
$$

where r is a position vector, and k_0 and k_r denote wave number vectors of the object
and reference waves, directing to the wave propagation. If the intensity $I(\omega)$ is proportional to the amplitude transmittance of the hologram, the reconstructed wave by
using the read out beam $T_{read}(\omega)\exp(ik_{read}r)$ is given by

$$
\begin{aligned}
T_{out}(\omega) &= I(\omega)T_{read}(\omega)\exp(ik_{read}r) \\
&= [|T_0(\omega)|^2 + |T_r(\omega)|^2]T_{read}(\omega)\exp(ik_{read}r) \\
&\quad + T_0(\omega)T_r^*(\omega)T_{read}^*(\omega)\exp[i(k_0 - k_r + k_{read})r] \\
&\quad + T_0^*(\omega)T_r(\omega)T_{read}(\omega)\exp[-i(k_0 - k_r - k_{read})r].
\end{aligned}
\tag{11.41}
$$

If $k_r = k_{read}$ and $T_r(\omega)$ and $T_{read}(\omega)$ have spectra with sufficiently wide and
flat as compared with that of $T_0(\omega)$, the second and third terms in Eq. (11.41) are
considered as

$$
T_0(\omega)\exp[ik_0r] + T_0^*(\omega)\exp[-i(k_0 - 2k_r)r]
\tag{11.42}
$$

and its temporal Fourier transform by the optical system shown in Fig. 11.11(b) gives

$$
T_{out}(t) \approx T_0(t)\exp[ik_0r] + T_0(-t)\exp[-i(k_0 - 2k_r)r].
\tag{11.43}
$$

The first term is the original pulse reconstructed and the second term is its time-
inversion pulse.

If $T_r(\omega)$ has the spectrum with sufficiently wide and flat as compared with that
of $T_0(\omega)$ and $T_{read}(\omega)$, the reconstructed beam includes $T_0(\omega)T_{read}(\omega)\exp[ik_0r] +$
$T_0^*(\omega)T_{read}(\omega)\exp[-i(k_0 - 2k_r)r]$, and its temporal Fourier transform gives the temporal correlation $T_0(t) * T_{read}(t)$, by which the temporal matched filtering for optical
pulses is realized.

The optical system shown in Fig. 11.9 is used in many measurements for ultra-
short pulses. Spatial Fourier transform of Eq. (11.42) gives the temporal wave shape
$T_0(t)$ as a spatial wave shape. The hologram intensity of Eq. (11.40) is used to measure the phase of the spectrum [6].

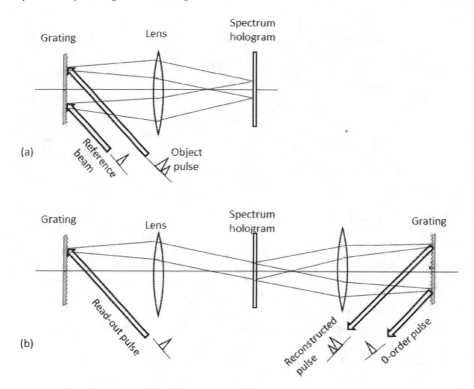

Figure 11.11   Optical system for spectral holography; (a) recording and (b) reconstruction.

If the phases of the object and reference beams are denoted by $\phi_0(\omega)$ and $\phi_r(\omega)$, respectively, Eq. (11.40) is rewritten as

$$I(\omega) = |T_0(\omega)|^2 + |T_r(\omega)|^2 + 2|T_0(\omega)||T_r(\omega)|\cos[\phi_0(\omega) - \phi_r(\omega) + (k_0 - k_r)r].$$
$$(11.44)$$

If the phase of the reference spectrum is known, the phase of the object $\phi_0(\omega)$ can be estimated [7].

REFERENCES

1. Martinez, O. E. 1986. Grating and prism compressors in the case of finite beam size. *J. Opt. Soc. Amer.* B3: 929.
2. Danailov, M. B. and Christov, I. P. 1989. Time-space shaping of light pulses by Fourier optical processing. *J. Mod. Opt.* 36: 725.
3. Yasuno, Y., Suto, Y., Yoshikawa, N., Itoh, M., Komori, K., Watanabe, M. and Yata-gaT. 2000. Time-space Conversion of Femtosecond light Pulse by Spatio-temporal Joint Transform Correlator. *Opt. Commun.* 177: 135.

4. Yasuno, Y., Nakamura, M., Suto, Y., Itoh, M. and Yatagai, T. 2000. Optical coherence tomography by spectral interferometric joint transform correlator. *Opt. Commun.* 186: 51.

5. Weiner, A. M., Leaird, D. E., Reize, D. H. and Paek, E. G. 1992. Femtosecond spectral holography. *IEEE J. Quantum. Electron.* 28: 2251.

6. Nuss, M. C., Li, M., Chiu, T. H., Weiner, A. M. and Partov, A. 1994. Time-to-space mapping of femtosecond pulses. *Opt. Lett.* 19: 664.

7. Meshulach, D., Yelin, D. and Silberberg, Y. 1997. Real-time spatial–spectral interference measurements of ultrashort optical pulses. *J. Opt. Soc. Amer.* B14: 2095.

12 Wigner Distribution Function

The Fourier transform analysis can describe the relation between a real signal domain and its spectral domain. The Wigner distribution function (WDF) is a 2-D function, which can represent the signal together in the real and frequency domains. Its projection to the real or frequency axis gives the power in a spectral or real domain. The WDF can describe some optical functions, such as the lens effect, the phase modulation, the diffraction grating, the wave propagation, and so on. Its applications to optical signal processing are presented. The four dimensional 4-D WDF in the space-time signal is also introduced.

12.1 WDF FOR SPATIAL SIGNAL

12.1.1 DEFINITION AND ITS PROPERTIES

The Wigner distribution function (WDF) is defined to represent conjugate physical quantities with uncertainty relation, like the position and the momentum in quantum mechanics. In this chapter, we consider the space position x and the spatial frequency v. The WDF of a real signal $f(x)$ is defined as [1, 2]

$$W_f(x, v) \equiv \int_{-\infty}^{\infty} f\left(x + \frac{x'}{2}\right) f^*\left(x - \frac{x'}{2}\right) \exp(-i2\pi x' v) dx'. \qquad (12.1)$$

The WDF is also defined as

$$W_f(x, v) \equiv \int_{-\infty}^{\infty} F\left(v + \frac{v'}{2}\right) F^*\left(v - \frac{v'}{2}\right) \exp(i2\pi x v') dv', \qquad (12.2)$$

due to the symmetrical property in the space and frequency domains, where $F(n)$ is the Fourier transform of $f(x)$.

Roughly speaking, the WDF gives the energy distribution in the real domain and frequency domain of a signal $f(x)$. It is obvious because of the following properties,

$$\int_{-\infty}^{\infty} W_f(x, v) dv = |f(x)|^2 \qquad (12.3)$$

$$\int_{-\infty}^{\infty} W_f(x, v) dx = |F(v)|^2 \qquad (12.4)$$

$$\iint_{-\infty}^{\infty} W_f(x, v) dx dv = \int |f(x)|^2 dx = \int |F(v)|^2 dv = \text{Total energy}, \qquad (12.5)$$

The WDFs of some common signals are given in Table 12.1. Figure 12.1 shows the WDF of $f(x) = \text{rect}(x)$, for example.

DOI: 10.1201/9781003121916-12

Table 12.1
WDFs of Common Signals

$f(x)$	$W_f(x,v)$
$\exp(i2\pi x v_0)$	$\delta(v-v_0)$
$\delta(x-x_0)$	$\delta(x-x_0)$
$\exp[i\pi(\alpha x^2 + 2\beta x + \gamma)]$	$\delta(v-\alpha x-\beta)$
$\text{rect}(x)$	$2(1-2|x|)\text{rect}(x)\text{sinc}[2(1-2|x|)v]$

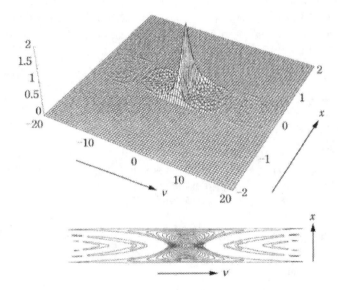

Figure 12.1   WDF of $\text{rect}(x)$.

In order to obtain an original function $f(x)$ from its WDF, at first, we have

$$f\left(x+\frac{x'}{2}\right)f^*\left(x-\frac{x'}{2}\right) = \int_{-\infty}^{\infty} W_f(x,v)\exp(i2\pi x'v)dv \qquad (12.6)$$

by inverse Fourier transforming Eq. (12.1). It should be noted that a given 2-D WDF can not always be decomposed like Eq. (12.6). If a 2-D function is known as $W_f(x,v)$, the following relation is valid

$$f(x) = \frac{1}{f^*(0)} \int_{-\infty}^{\infty} W_f(x/2,v)\exp(i2\pi xv)dv \qquad (12.7)$$

Table 12.2 shows some important properties of WDFs.

Table 12.2
Properties of WDFs

$W_f^*(x, v) = W_f(x, v)$	WDF is real function
WDF of $f(x - x_0)$	$W_f(x - x_0, v)$
WDF of $f(x) \exp(i2\pi v_0 x)$	$W_f(x, v - v_0)$
WDF of $f_1(x) * f_2(x)$	$\int W_{f_1}(x - x', v) W_{f_2}(x', v) dx'$
WDF of $f_1(x) \cdot f_2(x)$	$\int W_{f_1}(x, v - v') W_{f_2}(x, v') dv'$
WDF of $F(v) = \mathscr{F}[f(x)]$	$W_F(x, v) = W_f(-v, x)$

12.1.2   WDF IN OPTICAL SYSTEM

12.1.2.1   Lens Effect

The wave f_L after passing through a lens whose focal length l_F is given by

$$f_L(x) = f_0(x) \exp\left(-i\frac{\pi}{\lambda l_F} x^2\right) \tag{12.8}$$

according to Eq. (6.13), where $f_0(x)$ is an input wave. Similar to Eqs. (6.19) and (6.36), introducing a coordinate of μ in the Fourier plane, we have

$$v = \frac{\mu}{\lambda l}, \tag{12.9}$$

where l denotes a constant of an appropriate length or a focal length, and $l = l_F \eta$ with a constant η. The WDF of Eq. (12.8) is given by

$$
\begin{aligned}
W_{f_L}(x, \mu) &= \int f_L\left(x + \frac{x'}{2}\right) f_L^*\left(x - \frac{x'}{2}\right) \exp\left(-i2\pi \frac{\mu x'}{\lambda l}\right) dx' \\
&= \int f_0\left(x + \frac{x'}{2}\right) f_0^*\left(x - \frac{x'}{2}\right) \exp\left[-i2\pi \frac{x'}{\lambda l}(\mu + \eta x)\right] dx' \\
&= W_{f_0}(x, \mu + \eta x).
\end{aligned}
\tag{12.10}
$$

This means the lens effect shifts the WDF to the direction μ, which is proportional to the amount of x, as shown in Fig. 12.2.

12.1.2.2   Space Propagation

When the wave $f(x)$ propagates at a distance of $z = \xi l$, according to Eq. (6.2), we have

$$u(x) = f(x) * \exp\left(i\frac{\pi}{\lambda z} x^2\right). \tag{12.11}$$

Fourier transforming Eq. (12.11) and using Eq. (12.9), we have

$$U(\mu) = F(\mu) \exp\left(-i\pi \frac{\xi}{\lambda l} \mu^2\right) \tag{12.12}$$

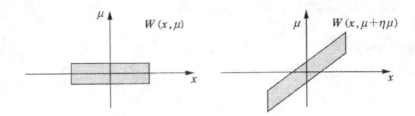

Figure 12.2   Change of WDF after lens.

WDF of Eq. (12.12) is given by

$$
\begin{aligned}
W_U(x,\mu) &= \int U\left(\mu + \frac{\mu'}{2}\right) U^*\left(\mu - \frac{\mu'}{2}\right) \exp\left(i2\pi\frac{x}{\lambda l}\mu'\right) d\mu' \\
&= \int F\left(\mu + \frac{\mu'}{2}\right) F^*\left(\mu - \frac{\mu'}{2}\right) \exp\left[i2\pi(x - \xi\mu)\frac{\mu'}{\lambda l}\right] d\mu' \\
&= \int f_0\left(x + \frac{x'}{2}\right) f_0^*\left(x - \frac{x'}{2}\right) \exp\left[-i2\pi(\mu - \xi x)\frac{x'}{\lambda l}\right] dx' \\
&= W_{f_0}(x - \xi\mu, \mu).
\end{aligned}
\tag{12.13}
$$

12.2   WDF FOR SPATIO-TEMPORAL SIGNAL

12.2.1   EXTENSION TO SPATIO-TEMPORAL SIGNALS

In Sec. 12.1, a two variable WDF for the space domain x and its frequency domain v was defined. Consider the extension to a four variable function for the space domain x and its frequency domain η and the time domain t and its angular frequency ω [3]. For the electric field $E(x,t)$, its Fourier transform in time domain $\tilde{E}(x,\omega)$ and its Fourier transform in space domain $\hat{E}(\zeta,t)$ are defined as

$$
\tilde{E}(x,\omega) = \int_{-\infty}^{\infty} E(x,t)\exp(i\omega t)dt
\tag{12.14}
$$

and

$$
\hat{E}(\zeta,t) = \frac{1}{2\pi}\int_{-\infty}^{\infty} E(x,t)\exp(-i\zeta x)dx,
\tag{12.15}
$$

where ω and ζ denote the frequency in the time domain and in the space domain, respectively.

Furthermore, the 2-D Fourier transform in the space and time domain is defined as

$$
\hat{\tilde{E}}(\zeta,\omega) = \frac{1}{2\pi}\iint_{-\infty}^{\infty} E(x,t)\exp[-i(\zeta x - \omega t)]dxdt.
\tag{12.16}
$$

The spatio-temporal WDF is defined as

$$
W^{ST}(x,\zeta,t,\omega)
$$
$$
=\frac{1}{2\pi}\iint_{-\infty}^{\infty} E\left(x+\frac{x'}{2},t+\frac{t'}{2}\right)E^*\left(x-\frac{x'}{2},t-\frac{t'}{2}\right)\exp[i(-\zeta x'+\omega t')]\mathrm{d}x'\mathrm{d}t'.
$$
$$(12.17)$$

The spatio-temporal WDFs a real function and a variety of definitions for variables x, ζ, t and ω are given by

$$
W^{ST}(x,\zeta,t,\omega)
$$
$$
=\frac{1}{(2\pi)^2}\iint_{-\infty}^{\infty} \tilde{E}\left(x+\frac{x'}{2},\omega+\frac{\omega'}{2}\right)\tilde{E}^*\left(x-\frac{x'}{2},\omega-\frac{\omega'}{2}\right)\exp[i(-\zeta x'-\omega' t)]\mathrm{d}x'\mathrm{d}\omega'
$$
$$(12.18)$$

$$
W^{ST}(x,\zeta,t,\omega)
$$
$$
=\iint_{-\infty}^{\infty} \hat{E}\left(\zeta+\frac{\zeta'}{2},t+\frac{t'}{2}\right)\hat{E}^*\left(\zeta-\frac{\zeta'}{2},t-\frac{t'}{2}\right)\exp[i(\zeta'x+\omega t')]\mathrm{d}\zeta'\mathrm{d}t' \quad (12.19)
$$

$$
W^{ST}(x,\zeta,t,\omega)
$$
$$
=\frac{1}{2\pi}\iint_{-\infty}^{\infty} \hat{\tilde{E}}\left(\zeta+\frac{\zeta'}{2},\omega+\frac{\omega'}{2}\right)\hat{\tilde{E}}^*\left(\zeta-\frac{\zeta'}{2},\omega-\frac{\omega'}{2}\right)\exp[i(\zeta'x-\omega' t)]\mathrm{d}\zeta'\mathrm{d}\omega'.
$$
$$(12.20)$$

The space WDFs $W^S(x,\zeta,t)$ and $W^S(x,\zeta,\omega)$ and the time WDFs $W^T(x,t,\omega)$ and $W^T(\zeta,t,\omega)$ are obtained from the spatio-temporal WDFs $W^{ST}(x,\zeta,t,\omega)$,

$$
W^S(x,\zeta,t)=\frac{1}{2\pi}\int_{-\infty}^{\infty} W^{ST}(x,\zeta,t,\omega)\mathrm{d}\omega \tag{12.21}
$$

$$
W^S(x,\zeta,\omega)=\int_{-\infty}^{\infty} W^{ST}(x,\zeta,t,\omega)\mathrm{d}t \tag{12.22}
$$

$$
W^T(x,t,\omega)=\int_{-\infty}^{\infty} W^{ST}(x,\zeta,t,\omega)\mathrm{d}\zeta \tag{12.23}
$$

$$
W^T(\zeta,t,\omega)=\frac{1}{2\pi}\int_{-\infty}^{\infty} W^{ST}(x,\zeta,t,\omega)\mathrm{d}x. \tag{12.24}
$$

Similar to Eqs. (12.3) and (12.4), integrations of the spatio-temporal WDF by arbitrary two variables give the energy distributions for the other variables,

$$
\frac{1}{2\pi}\iint_{-\infty}^{\infty} W^{ST}(x,\zeta,t,\omega)\mathrm{d}\zeta\mathrm{d}\omega=|E(x,t)|^2 \tag{12.25}
$$

$$
\iint_{-\infty}^{\infty} W^{ST}(x,\zeta,t,\omega)\mathrm{d}\zeta\mathrm{d}t=|\tilde{E}(x,\omega)|^2 \tag{12.26}
$$

$$\frac{1}{(2\pi)^2} \iint_{-\infty}^{\infty} W^{ST}(x,\zeta,t,\omega)\mathrm{d}x\mathrm{d}\omega = |\hat{E}(\zeta,t)|^2 \qquad (12.27)$$

$$\frac{1}{2\pi} \iint_{-\infty}^{\infty} W^{ST}(x,\zeta,t,\omega)\mathrm{d}x\mathrm{d}t = |\hat{\tilde{E}}(\zeta,\omega)|^2 \qquad (12.28)$$

$$\frac{1}{2\pi} \iint_{-\infty}^{\infty} W^{ST}(x,\zeta,t,\omega)\mathrm{d}t\mathrm{d}\omega = \int W^S(x,\zeta,t)\mathrm{d}t \qquad (12.29)$$

$$\frac{1}{2\pi} \iint_{-\infty}^{\infty} W^{ST}(x,\zeta,t,\omega)\mathrm{d}x\mathrm{d}\zeta = \int W^T(\zeta,t,\omega)\mathrm{d}\zeta. \qquad (12.30)$$

Finally, the integration of the spatio-temporal WDF by all four variables gives the total energy,

$$\frac{1}{(2\pi)^2} \iint_{-\infty}^{\infty} W^{ST}(x,\zeta,t,\omega)\mathrm{d}x\mathrm{d}\zeta\mathrm{d}t\mathrm{d}\omega = Eng. \qquad (12.31)$$

12.2.2 LENS EFFECT IN SPATIO-TEMPORAL WDF

Because the change of the spatial component of WDF after passing a lens is given by Eq. (12.10), the spatio-temporal WDF is represented by [4]

$$W_{out}^{ST}(x,\zeta,t,\omega) = W_{in}^{ST}\left(x,\zeta+\frac{k_0}{f}x,t,\omega\right), \qquad (12.32)$$

where f and k_0 denote the focal length of the lens and the wave number for the central wavelength, respectively, and

$$\zeta = 2\pi\nu. \qquad (12.33)$$

Generally, consider how the WDF is changed by the phase modulation. After passing through a lens, the phase of the wave is modulated

$$E_{out}(x,t) = E_{in}(x,t)\exp\left(-\mathrm{i}\frac{k_0}{2f}x^2\right) \qquad (12.34)$$

according to Eq. (12.8). If the phase modulation is given by

$$\Phi(x,\zeta,t,\omega) = -\frac{k_0}{2f}x^2. \qquad (12.35)$$

Then substituting Eqs. (12.34) and (12.35) into Eq. (12.16), we have

$$W_{out}^{ST}(x,\zeta,t,\omega) = \frac{1}{2\pi}\iint_{-\infty}^{\infty} E_{in}\left(x+\frac{x'}{2},t+\frac{t'}{2}\right)E_{in}^*\left(x-\frac{x'}{2},t-\frac{t'}{2}\right)$$
$$\times \exp\left\{\mathrm{i}\left[\Phi\left(x+\frac{x'}{2}\right)-\Phi\left(x-\frac{x'}{2}\right)\right]\right\}\exp[\mathrm{i}(-\zeta x'+\omega t')]\mathrm{d}x'\mathrm{d}t'. \qquad (12.36)$$

By expanding $\Phi(x)$, we have

$$W_{out}^{ST}(x,\zeta,t,\omega) = \frac{1}{2\pi}\iint_{-\infty}^{\infty} E_{in}\left(x+\frac{x'}{2},t+\frac{t'}{2}\right)E_{in}^*\left(x-\frac{x'}{2},t-\frac{t'}{2}\right)$$
$$\times \exp\{-\mathrm{i}[\zeta-\Phi'(x)]x'\}\exp(-\mathrm{i}\omega t')\mathrm{d}x'\mathrm{d}t'. \qquad (12.37)$$

Finally, we have

$$W_{out}^{ST}(x,\zeta,t,\omega) = W_{in}^{ST}\left[x,\zeta - \frac{\partial\Phi(x,\zeta,t,\omega)}{\partial x},t,\omega\right]. \qquad (12.38)$$

By the spatial phase modulation, the spatial frequency of WDF is changed as described by Eq. (12.38). The phase modulation of a lens given by Eq. (12.35) changes WDF as

$$W_{out}^{ST}(x,\zeta,t,\omega) = W_{in}^{ST}\left(x,\zeta + \frac{k_0}{f}x,t,\omega\right). \qquad (12.39)$$

In the case of the temporal phase modulation, the frequency is changed similarly.

12.2.3   TEMPORAL PHASE MODULATOR (TIME LENS)

If the input wave $E_{in}(x,t)$ is sinusoidally modulated in the phase, the output wave is given by [5, 6]

$$E_{out}(x,t) = E_{in}(x,t)\exp(-i\Phi_m\cos(\omega_m t)), \qquad (12.40)$$

where Φ and ω_m denote the amplitude and the angular frequency of the phase modulation. If the time considered is shorter than the modulation time, the sinusoidal modulation is approximated by terms up to the second order and then we have

$$E_{out}(x,t) = E_{in}(x,t)\exp\left[-i\Phi_m\left(1 - \frac{\omega_m^2 t^2}{2}\right)\right]. \qquad (12.41)$$

Since this phase modulation is given by

$$\Phi(x,\zeta,t,\omega) = -\Phi_m\left(1 - \frac{\omega_m^2 t^2}{2}\right) \qquad (12.42)$$

and its partial derivative is given by

$$\frac{\partial\Phi}{\partial t} = \Phi_m\omega_m^2 t \qquad (12.43)$$

then the input and output relation of the spatio-temporal WDFs for the temporal phase modulator is given by

$$W_{out}^{ST}(x,\zeta,t,\omega) = W_{in}^{ST}(x,\zeta,t,\omega + \Phi_m\omega_m^2 t). \qquad (12.44)$$

Comparing with Eqs. (12.39) and (12.44), the phase modulation of Eq. (12.41) operates the lens effect in the temporal domain and then this is called the time lens.

12.2.4   PROPAGATION AND DISPERSION

In the dispersive media, the Fresnel diffraction and the time-frequency dispersion are caused simultaneously. The paraxial wave after passing through the dispersive medium is represented by

$$\hat{\hat{E}}(\zeta,\omega,z) = \hat{\hat{E}}(\zeta,\omega,0)\exp[ik(\omega)z]\exp\left[-i\frac{\zeta^2}{2k(\omega)}z\right], \qquad (12.45)$$

where $k(\omega)$ denotes the propagation constant with the refractive index $n(\omega)$ and $k(\omega) = \omega n(\omega)/c$. If the frequency width is sufficiently small compared to the central frequency ω_m, the following approximation is valid,

$$k(\omega) = k_0 + k_0'\omega + \frac{1}{2}k_0''\omega^2, \tag{12.46}$$

where k_0' and k_0'' denote the first and second derivative of $k(\omega)$, respectively. Equation (12.45) is rewritten as

$$\hat{E}(\zeta,\omega,z) = \hat{E}(\zeta,\omega,0)\exp\left[i\left(k_0 + k_0'\omega + \frac{1}{2}k_0''\omega^2\right)z\right]\exp\left(-i\frac{\zeta^2}{2k_0}z\right). \tag{12.47}$$

This means that the effect of the propagation and dispersion causes the phase modulation

$$\Phi(x,\zeta,t,\omega) = k_0 z + k_0'\omega z + \frac{1}{2}k_0''\omega^2 z - \frac{\zeta^2 z}{2k_0} \tag{12.48}$$

to the wave. The derivatives of Eq. (12.48) are

$$\frac{\partial \Phi}{\partial \zeta} = -\frac{z}{k_0}\zeta \tag{12.49}$$

$$\frac{\partial \Phi}{\partial \omega} = k_0'z + k_0''z\omega. \tag{12.50}$$

The input–output relation of the spatio-temporal WDF in the propagation of the wave and the dispersion of the medium is given by

$$W_{out}^{ST}(x,\zeta,t,\omega) = W_{in}^{ST}\left(x - \frac{z}{k_0}\zeta, \zeta, t - k_0'z - k_0''z\omega, \omega\right). \tag{12.51}$$

When Eq. (12.51) is compared with Eq. (12.38), it should be noted that the sign of ζ is "−" in the first component of the term in the right-hand side, like $x - z\zeta/k_0$. This is because signs of the Fourier integral kernels in Eqs. (12.16) and (12.20) are opposite. Due to the same reason, the right-hand side in Eq. (12.44) is $\omega + \Phi_m\omega_m^2 t$.

12.2.5 DIFFRACTION GRATING

The wave after diffraction from a grating is given by

$$E_{out}(x,t) = E_{in}(\alpha x, t - \beta x) \tag{12.52}$$

as described in Sec. 11.1, where α and β are constants

$$\alpha = \frac{\cos\theta_i}{\cos\theta_d} \tag{12.53}$$

$$\beta = \frac{2\pi p}{d\omega_0\cos\theta_d}, \tag{12.54}$$

where p is the diffraction order of the grating, θ_i and θ_d denote the incidence angle and the diffraction angle, respectively, and d denotes the grating pitch. Submitting Eq. (12.52) into Eq. (12.16) and spatio-temporal Fourier transforming, we have

$$\hat{E}_{out}(\zeta,\omega) = E_{in}\left(\frac{\zeta}{\alpha} - \frac{\beta}{\alpha}\omega, \omega\right) \tag{12.55}$$

The effect of the grating in the spatio-temporal WDF is

$$W_{out}^{ST}(x,\zeta,t,\omega) = \frac{1}{\alpha}W_{in}^{ST}\left(\alpha x, \frac{\zeta}{\alpha} - \frac{\beta}{\alpha}\omega, t - \beta x, \omega\right). \tag{12.56}$$

12.2.6 MATRIX REPRESENTATION OF WDF TRANSFORMATION

The discussion on the transform of the spatio-temporal WDF shows that the transform can be performed by a linear transform of variables. Here we introduce an input vector X_i and an output vector X_0 with four elements (x,ζ,t,ω) and a 4×4 matrix $\mathbf{M}$ [3]. The input-output system is described by the spatio-temporal WDF

$$W_{out}^{ST}(X_0) = mW_{in}^{ST}(\mathbf{M} \cdot X_i), \tag{12.57}$$

where m is a system constant. The input-output variables are transformed by

$$X_0 = \mathbf{M} \cdot X_i = \begin{pmatrix} x_0 \\ \zeta_0 \\ t_0 \\ \omega_0 \end{pmatrix} = \begin{pmatrix} A_{xx} & A_{x\zeta} & A_{xt} & A_{x\omega} \\ A_{\zeta x} & A_{\zeta\zeta} & A_{\zeta t} & A_{\zeta\omega} \\ A_{tx} & A_{t\zeta} & A_{tt} & A_{t\omega} \\ A_{\omega x} & A_{\omega\zeta} & A_{\omega t} & A_{\omega\omega} \end{pmatrix} \begin{pmatrix} x_i \\ \zeta_i \\ t_i \\ \omega_i \end{pmatrix}. \tag{12.58}$$

If n systems are connected in series, the input-output relation of WDF is represented by

$$W_{out}^{ST}(X_0) = m_n m_{n-1} \cdots m_1 W_{in}^{ST}(\mathbf{M}_n \cdot \mathbf{M}_{n-1} \cdots \mathbf{M}_1 \cdot X_i). \tag{12.59}$$

12.2.6.1 Lens

$$\mathbf{M}_{lens} = \begin{pmatrix} 1 & 0 & 0 & 0 \\ k_0/f & 1 & 0 & 0 \\ 0 & 0 & 1 & 0 \\ 0 & 0 & 0 & 1 \end{pmatrix}, \tag{12.60}$$

where k_0 denotes a constant of wave number given by $k_0 = 2\pi/\lambda_0$ and f denotes the focal length of a lens.

12.2.6.2 Temporal Phase Modulation (Time Lens)

$$\mathbf{M}_{pmod} = \begin{pmatrix} 1 & 0 & 0 & 0 \\ 0 & 1 & 0 & 0 \\ 0 & 0 & 1 & 0 \\ 0 & 0 & \Phi_m\omega_m^2 & 1 \end{pmatrix}, \tag{12.61}$$

where Φ_m denotes the amplitude of a sinusoidal phase modulation to perform a time lens and ω_m its angular frequency.

12.2.6.3 Propagation and Dispersion

$$
\mathbf{M}_{prop} = \begin{pmatrix} 1 & -z/k_0 & 0 & 0 \\ 0 & 1 & 0 & 0 \\ 0 & 0 & 1 & -k_0'' z \\ 0 & 0 & 0 & 1 \end{pmatrix}, \tag{12.62}
$$

where z and k_0'' denote the propagation distance and the second derivative (group velocity) of $k(\omega)$, respectively.

12.2.6.4 Grating

$$
\mathbf{M}_{grating} = \begin{pmatrix} \alpha & 0 & 0 & 0 \\ 0 & 1/\alpha & 0 & -\beta/\alpha \\ -\beta & 0 & 1 & 0 \\ 0 & 0 & 0 & 1 \end{pmatrix}, \tag{12.63}
$$

where $\alpha = \cos\theta_i / \cos\theta_d$ and $\beta = 2\pi p/(d\omega_0 \cos\theta_d)$, θ_i and θ_d denote the incident angle and the diffraction angle, p the diffraction order, d the grating pitch, and ω_0 the central frequency of light.

As described above, to consider a general system, the product of matrices gives the matrix of the total system. As an example, the spatial Fourier transform system with a 2-f system, which consists of propagation, lens, and propagation, is given by

$$
\mathbf{M} = \mathbf{M}_{prop}\mathbf{M}_{lens}\mathbf{M}_{prop} = \begin{pmatrix} 0 & -f/k_0 & 0 & 0 \\ f/k_0 & 0 & 0 & 0 \\ 0 & 0 & 1 & 0 \\ 0 & 0 & 0 & 1 \end{pmatrix}, \tag{12.64}
$$

where $k_0'' = 0$ is assumed. This means the system exchanges the relationship between x and ζ. If x and ζ are represented in orthogonal coordinates, the spatial Fourier transform corresponds to the 90-degree rotation of the coordinate system, as described in Sec. 13.2.

The system matrix of a spectroscopic optical system consisting of a grating-lens pair is given by

$$
\mathbf{M} = \mathbf{M}_{lens}\mathbf{M}_{prop}\mathbf{M}_{grating} = \begin{pmatrix} \alpha & -f/(\alpha k_0) & 0 & \beta f/(\alpha k_0) \\ \alpha k_0/f & 0 & 0 & 0 \\ -\beta & 0 & 1 & 0 \\ 0 & 0 & 0 & 1 \end{pmatrix}. \tag{12.65}
$$

The applications of this system are discussed in Sec. 11.1.

REFERENCES

1. Classen, T. A. C. M. and Mecklenbrauker, W. F. G. 1980. The Wigner distribution-a tool for time-frequency signal analysis. *Philips J. Res.* 35: 217–250, 276–300, 372–389.
2. Lohmann, A. W. and Soffer, B. H. 1994. Relationships between the Radon–Wigner and fractional Fourier transforms. *J. Opt. Soc. Amer.* A11: 1798.

3. Paye, J. and Migus, A. 1995. Space–time Wigner functions and their application to the analysis of a pulse shaper. *J. Opt. Soc. Amer.* B12: 1480.
4. Bastiaans, M. J. 1978. The Wigner distribution function applied to optical signals and systems. *Opt. Commun.* 25: 26.
5. Kolner, B. H. and Nazarathy, M. 1989. Temporal imaging with a time lens. *Opt. Lett.* 14: 630.
6. Godil, A. A., Auld, B. A. and Bloom, D. M. 1994. Picosecond time-lenses. *IEEE J. Quantum Electron.* 30: 827.

13 Fractional Fourier Transform

A generalization of the conventional Fourier transform is called the fractional Fourier transform, which is defined by the Wigner distribution function (WDF) gives the Fourier transform. The rotation of WDF can be performed by a combination of shifts of WDF. Based on this sequential shift operations, optical systems performing the fractional Fourier transform are presented. Some applications of optical computing, such as the Wiener filtering, optical correlator, matched filter, and joint Fourier transform correlator are presented. A signal and noise separation method using multiple fractional Fourier transforms is discussed.

13.1 DEFINITION OF FRACTIONAL FOURIER TRANSFORM

Fourier transform of a signal $f(x)$ is defined by

$$F(v) = \mathscr{F}[f(x)] = \int_{-\infty}^{\infty} f(x)\exp(-i2\pi xv)dx \tag{13.1}$$

Distribution Function and its inverse Fourier transform is

$$f(x) = \mathscr{F}^{-1}[F(v)] = \int_{-\infty}^{\infty} F(v)\exp(i2\pi xv)dv. \tag{13.2}$$

By introducing Fourier transform operator $\mathscr{F}^a$, Fourier transform is represented in $a = 1$ and the inverse Fourier transform in $a = -1$, because Eqs. (13.1) and (13.2) are valid. In the case of $a = 0$,

$$\mathscr{F}^0 = \mathscr{F}^{-1}[\mathscr{F}^1[f(x)]] = f(x), \tag{13.3}$$

which corresponds to no transforming. In the case of $a = 2$,

$$\mathscr{F}^2 = \mathscr{F}^1[\mathscr{F}^1[f(x)]] = f(-x). \tag{13.4}$$

In general, in the case when a is not integral, the transform is called the fractional Fourier transform [1, 2].

13.2 SOME REPRESENTATIONS OF FRACTIONAL FOURIER TRANSFORM

According to Table 12.2, the WDF of Fourier transform $F(v)$ of a signal $f(x)$ is given by

$$W_F(x,v) = \int F\left(x + \frac{x'}{2}\right)F^*\left(x - \frac{x'}{2}\right)\exp(-i2\pi x'v)dx' \tag{13.5}$$

DOI: 10.1201/9781003121916-13

197

and

$$W_F(x, v) = W_f(-v, x).$$ (13.6)

This means that the Fourier transform rotates its WDF by 90° clockwise, as shown in Fig. 13.1. In general, the a-th order fractional Fourier transform rotates the coordinate axis for its WDF by an angle

$$\phi = \frac{\pi}{2}a.$$ (13.7)

Therefore, the WDF of the a-th order fractional Fourier transform is represented by

$$W_{F^a}(x, v) = W_f(x\cos\phi - v\sin\phi, x\sin\phi + v\cos\phi).$$ (13.8)

The fractional Fourier transform $\mathscr{F}^a[f(x)]$ is obtained by using Eqs. (13.7) and (13.8).

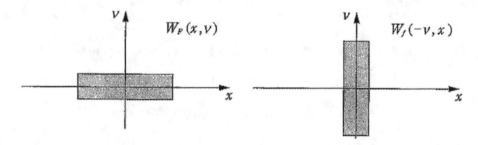

Figure 13.1 Rotation of WDF by Fourier transform.

Next, we show that the rotation of WDF can be performed by a combination of the shift of Eqs. (12.10) and (12.13) [3]. According to Fig. 13.2, sequential shifts of the WDF to the y-direction, the x-direction and the y-direction are made. The coordinates (x_0, y_0) of a point are transformed sequentially as

$$(x_0, y_0) \longrightarrow (x_0, y_0 + Ax_0) = (x_1, y_1)$$ (13.9)

$$(x_1, y_1) \longrightarrow (x_1 - By_1, y_1) = (x_2, y_2)$$ (13.10)

$$(x_2, y_2) \longrightarrow (x_2, y_2 + Cx_2) = (x_3, y_3)$$ (13.11)

and finally, we have

$$x_3 = x_0(1 - AB) - By_0$$ (13.12)

$$y_3 = y_0(1 - BC) + x_0(A + C - ABC).$$ (13.13)

In order that the sequential shifts become to the ϕ degrees rotation, it is necessary that

$$B = \sin\phi, \quad A = C = \tan\left(\frac{\phi}{2}\right).$$ (13.14)

Therefore,

$$\xi = \sin\phi, \quad \eta = \tan\left(\frac{\phi}{2}\right). \tag{13.15}$$

This procedure corresponds to the following optical operations, that is, at first, passing a lens with the focal length $l_F = 1/\eta$, then propagation by a distance $z = \xi l$ and again passing a lens with the focal length $l_F = 1/\eta$. The optical system of this operation is shown in Fig. 13.3(a). This is the optical system, which operates the a-th order fractional Fourier transform of a signal $u_0(x_0)$.

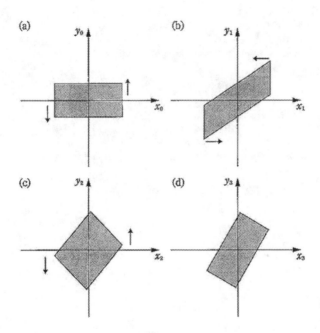

Figure 13.2   Rotation of the WDF by shifts.

Next, consider the representation of the a-th order fractional Fourier transform of a signal $u_{(x_0)}$. At first, the object $u_0(x_0)$ passes through a lens with the focal length $l_F = 1/\eta$.

$$u_1(x_0) = u_0(x_0)\exp\left(-i\frac{\pi\eta}{\lambda l}x_0^2\right). \tag{13.16}$$

Then, the Fresnel diffraction of a distance $z = \xi l$ is made

$$u_2(x_2) = u_1(x_2) * \exp\left(i\frac{\pi}{\lambda l\xi}x_2^2\right). \tag{13.17}$$

Its spectrum representation is

$$\tilde{u}_2(v_2) = \tilde{u}_1(v_2)\exp(-i\pi\lambda l\xi v_2^2), \tag{13.18}$$

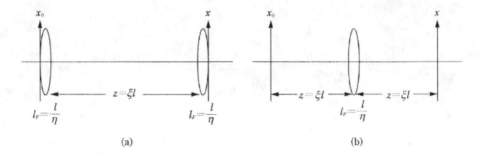

Figure 13.3 Optical system for fractional Fourier transform.

where

$$\tilde{u}_1(v_2) = \int u_1(x_2)\exp\left(-i\frac{2\pi}{\lambda l}x_2 v_2\right)dx_2. \tag{13.19}$$

After passing through a lens with the focal length $l_F = l/\eta$ again,

$$u_a(x) = u_2(x)\exp\left(-i\frac{\pi\eta}{\lambda l}x^2\right). \tag{13.20}$$

Finally, Eqs. (13.16), (13.17) and (13.20) and the transformation of Eq. (13.15) give the a-th order Fourier transform

$$u_a(x) = \mathscr{F}^a[u_0(x_0)]$$
$$= \int u_0(x_0)\exp\left[i\frac{\pi}{\lambda l\tan\phi}(x_0^2 + x^2)\right]\exp\left(-i\frac{2\pi}{\lambda l\sin\phi}xx_0\right)dx_0. \tag{13.21}$$

The rotation of the WDF by shift can be also performed by serial shifts to the x-, y- and x-directions. In this case,

$$\xi = \tan\left(\frac{\phi}{2}\right), \qquad \eta = \sin\phi. \tag{13.22}$$

The corresponding optical system is shown in Fig. 13.3(b).

From the discussion above and the mathematical requirement, the fractional Fourier transform of the a-th order $f_a(x)$ for a function $f(x)$ is defined again as

$$f_a(x) \equiv \mathscr{F}[f] = \int_{-\infty}^{\infty} K_a(x,x')f(x')dx', \tag{13.23}$$

where

$$K_a(x,x') = A_a\exp\left\{i\pi\left[(\cot\phi)x^2 - 2(\csc\phi)xx' + (\cot\phi)x'^2\right]\right\} \tag{13.24}$$

$$A_a = \sqrt{1 - i\cot\phi}. \tag{13.25}$$

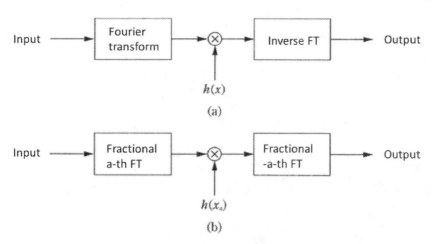

(a)

(b)

Figure 13.4 (a) Filtering by Fourier transform and (b) fractional Fourier transform.

13.3 APPLICATIONS TO OPTICAL COMPUTING

13.3.1 WIENER FILTERING

As mentioned in Sec. 8.3, the Wiener filter is optimum to restore the degraded image
with additive noise, described by Eq. (8.36), as shown in Fig. 13.4(a). The Wiener
filter is not always effective for the additive noise with WDF as shown in Fig. 13.5,
because the power spectrum of the degraded signal, which is given by the projection
of the WDF into the v-axis, Eq. (12.4), and the power spectrums of the signal and
the noise are overlapped.

To reduce the overlapping area in the spectral domain, the use of the fractional
Fourier transform of an appropriate order is effective [4, 5, 6]. In WDF with the x_a-
axis, the power spectra, which is given by the projection of WDF into the x_a-axis is
not overlapped. This means that the noise is easily eliminated using an appropriate
mask. The filtering procedure by the fractional Fourier transform is shown in Fig.
13.4(b). Its optimum filter corresponding to Eq. (8.36) is given by

$$h(x_a) = \frac{\iint K_a(x_a,x)K_{-a}(x_a,x')\phi_{fg}(x,x')dxdx'}{\iint K_a(x_a,x)K_{-a}(x_a,x')\phi_{gg}(x,x')dxdx'}, \tag{13.26}$$

where $K_a(x_a,x)$ is defined by Eq. (13.24), and $\phi_{fg}(x,x')$ and $\phi_{gg}(x,x')$ denote the
cross-correlation function between $f(x)$ and $g(x)$ and the autocorrelation of $g(x)$,
respectively.

For the signal with the noise as shown in Fig. 13.6, the noise reduction can be performed by multiple applications of the fractional Fourier transforms of appropriate
orders and cascaded optimum Wiener filtering.

Consider a signal

$$f(x) = \exp\left[-\pi(x-4)^2\right] \tag{13.27}$$

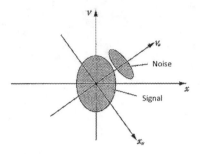

Figure 13.5   Wigner distribution of signal and noise.

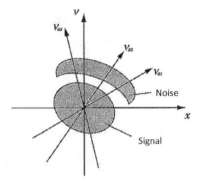

Figure 13.6   Separation of signal and noise by multiple fractional Fourier transforms.

as shown in Fig. 13.7(a), and the chirp noise

$$n(x) = \exp(-i\pi x^2)\mathrm{rect}\left(\frac{x}{16}\right). \tag{13.28}$$

The fractional Fourier transform of the order $a = 0.5$ is shown in Fig. 13.7(b), the result of masking as shown in Fig. 13.7(c), and finally, we have the noise reduced signal by the inverse fractional Fourier transform as shown in Fig. 13.7(d). For reference, the WDF of the degraded signal is shown in Fig. 13.8.

In general, the filtering by the fractional Fourier transform is effective in some cases for the signal with a chirp distorted noise.

13.3.2   CORRELATOR AND MATCHED FILTER

The correlation operation is defined by

$$g(x) = f(x) \star h^*(x) \tag{13.29}$$

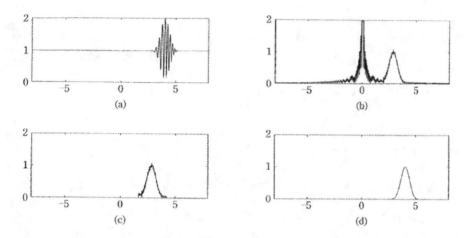

Figure 13.7 Noise reduction by fractional Fourier transform [2].

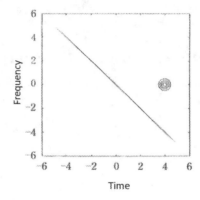

Figure 13.8 Wigner distribution function of the degraded signal of Fig. 13.7(a) [2].

and its Fourier transform is

$$G(v) = F(v) \cdot H^*(v). \tag{13.30}$$

It is reasonable to define the fractional correlation as

$$g_a(x_a) = f_a(x_a)h_a^*(x_a). \tag{13.31}$$

In the case $a = 0$, the fractional correlation corresponds to the product $g(x) = f(x)h^*(x)$ and in the case $a = 1$, the conventional correlation $G(v) = F(v)H^*(v)$.

In more general, the fractional correlation is defined as

$$g_{a_2}(x_a) = f_{a_1}(x_a)h_{a_1}^*(x_a). \tag{13.32}$$

This means that the fractional Fourier transforms of the order a_1 for $f(x)$ and $g(x)$, then the product of both transforms and finally the fractional inverse Fourier transform of the order a_2 for the product are calculated. This procedure can be described by

$$g(x_a) = \iint K_{fc}(x_a, x_a', x_a'') f(x_a') h_a^*(x_a'') \mathrm{d}x_a' \mathrm{d}x_a'', \qquad (13.33)$$

where

$$K_{fc}(x_a, x_a', x_a'') = \int K_{-a_2}(x_a, x_a''') K_{a_1}(x_a''', x_a') K_{-a_1}(x_a''', x_a'') \mathrm{d}x_a'''. \qquad (13.34)$$

It should be noted that the fractional correlation is not shift-invariant for $f(x)$, that is, the correlation depends on the position of the input signal. This is a big difference between the fractional correlation and the conventional correlation based on the Fourier transform.

An example of signal detection using fractional Fourier transform is shown in Fig. 13.9. Figure 13.9(a) shows an object to be detected, (b): its conventional autocorrelation, (c): the fractional autocorrelation of the order $a = 0.5$, and (d): the fractional autocorrelation of the order $a = 0.5$ in the case of a shifted object. The fractional correlation is not shift-invariant.

13.3.3   JOINT FRACTIONAL FOURIER TRANSFORM CORRELATOR

Figure 13.10 shows a block diagram of a joint fractional Fourier transform correlator [8]. If $p_1 = p_2 = 1$ and $p_3 = -1$, it is the conventional joint transform correlator.

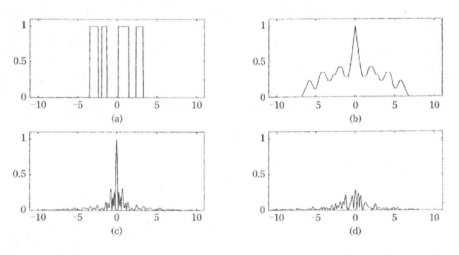

Figure 13.9 Matched filtering by fractional Fourier transform. (a) Object, (b) conventional autocorrelation, (c) fractional autocorrelation in $a = 0.5$ and (d) fractional autocorrelation in $a = 0.5$ for the object shifted [2].

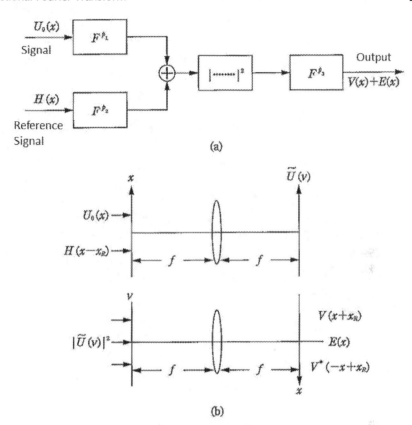

Figure 13.10 Joint transform correlator by fractional Fourier transform. (a) Block diagram, and (b) optical system for joint fractional Fourier transform correlator. $V(x)$: output signal and $E(x)$: signal except for output signal.

Let us discuss the joint transform correlator by using the WDF. As described in Sec. 8.7, the process is divided in two processes [8]. In the first process, an object signal $U_0(x)$ and a reference signal $H(x)$ are positioned in the same plane and Fourier transformed. The signals in the input plane are described by

$$U(x) = U_0(x) + H(x - x_R), \tag{13.35}$$

where x_R denotes a separation between the object and reference positions. The WDF representation of the signal $U(x)$ is shown in Fig. 13.11, where W_0 and W_H denotes WDFs of $U_0(x)$ and $H(x)$, respectively, and Δx_0 and Δx_H the extents of $U_0(x)$ and $H(x)$, respectively. Its Fourier transform is given by

$$\tilde{U}(v) = \tilde{U}_0(v) + \tilde{H}(v)\exp(-i2\pi v x_R) \tag{13.36}$$

and its WDF is shown in Fig. 13.12. WDFs of the object and reference signals locates in the same x area. This means that $\tilde{U}_0(x)$ and $\tilde{H}\exp(-i2\pi x x_R)$ can interact each

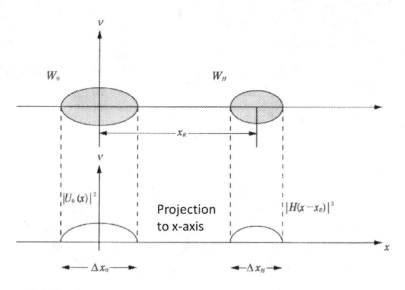

Figure 13.11   WDF of input $U(x)$.

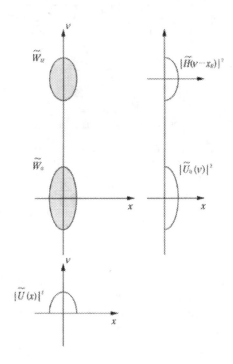

Figure 13.12   WDF of Eq. (13.36).

other. The projection of WDF to the x axis gives $\tilde{U}(x)$ and corresponds to the input in Fig. 13.10(b).

$$\int \tilde{W}(x,v)\mathrm{d}v = |\tilde{U}(x)|^2 = |\tilde{U}_0 + \tilde{H}\exp(-i2\pi xx_R)|^2$$

$$= |\tilde{U}_0(x)|^2 + |\tilde{H}(x)|^2 + \tilde{U}_0\tilde{H}^*\exp(i2\pi xx_R) + \tilde{U}_0^*\tilde{H}\exp(-i2\pi xx_R)$$

$$= \tilde{E}(x) + \tilde{V}(x)\exp(i2\pi xx_R) + \tilde{V}^*(x)\exp(-i2\pi xx_R), \tag{13.37}$$

where

$$\tilde{V}(x) = \tilde{U}_0(x)\tilde{H}^*(x) \tag{13.38}$$

and

$$\tilde{E}(x) = \left|\tilde{U}_0(x)\right|^2 + \left|\tilde{H}(x)\right|^2. \tag{13.39}$$

$\left|\tilde{U}(x)\right|^2$ is shown in Fig. 13.13, which consists of three parts. To obtain the correlation signal U_0H^*, the condition

$$x_R - \frac{\Delta x_H + \Delta x_0}{2} \geq \Delta x_0 \tag{13.40}$$

should be satisfied.

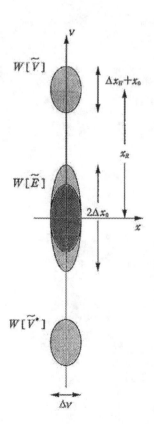

Figure 13.13 WDF of Eq. (13.37).

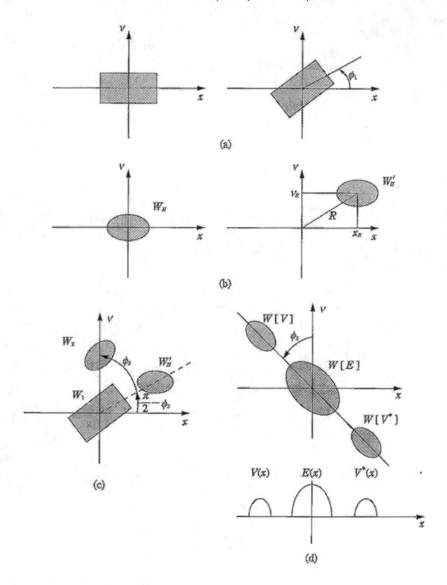

Figure 13.14   WDF of fractional Fourier transform joint correlator.

Consider the WDF representation of the joint fractional Fourier transform correlator. At first, the fractional Fourier transform of the p_1-th order of the signal $U(x)$ is calculated, which corresponds to the rotation of WDF by the angle of $\phi_1 = \frac{\pi}{2}p_1$, as shown in Fig. 13.14(a). In order to calculate $|U(x)|^2$, similar to the case of Eq.

(13.37), the reference signal is modulated by shift and linear phase modulation as

$$H(x) \rightarrow H(x - x_R) \exp(i2\pi x v_R) \qquad (13.41)$$

and then the fractional Fourier transform of the p_1-th order is calculated. This process is represented as

$$W_H(x, v) \rightarrow W_H(x - x_R, v - v_R) \qquad (13.42)$$

as shown in Fig. 13.14(b). Next, the fractional Fourier transform of the p_2-th order of this signal is calculated, which is obtained by rotation WDF by the angle $\phi_2 = \frac{\pi}{2} p_2$. To locates W_H' on the v-axis, the conditions

$$x_R = R\cos\left(\frac{\pi}{2} - \phi_2\right) = R\sin\phi_2 \qquad (13.43)$$

and

$$v_R = R\sin\left(\frac{\pi}{2} - \phi_2\right) = R\cos\phi_2 \qquad (13.44)$$

are necessary. Then the signal $|U(x)|^2$, corresponding to Fig. 13.11, is calculated, as shown in Fig. 13.14(c). Finally, the fractional Fourier transform of the p_2-th order of this signal is calculated by rotating the WDF by the angle $\phi_3 = \frac{\pi}{2} p_3$, as shown in Fig. 13.14(d). The separation distance of the correlation signal is

$$R = \sqrt{x_R^2 + v_R^2}. \qquad (13.45)$$

The final correlation signal is the projection of the WDF to the x-axis. The separation distance R does not depend on only the separation distance x_R but the spectral extent v_R, while the separation distance depends only on the distance between the signal and the reference signal x_R in the conventional joint transform correlator.

REFERENCES

1. Lohmann, A. W., Mendlovic, D. and Zalevsky, Z. 1998. Fractional Transforms in Optics, *Progress in Optics*. XXXVIII: 263–342.
2. Ozaktas, H. M., Zalevsky, Z. and Kutay, M. A. 2001. *The Fractional Fourier Transform*. John Wiley & Sons, Chichester.
3. Lohmann, A. W. 1993. Image rotation, Wigner rotation, and the fractional Fourier transform. *J. Opt. Soc. Amer.* A10: 2181.
4. Mendlovc, D. Zalevsky, Z. and Ozaktas, H. M. Applications of the fractional Fourier transform to optical pattern recognition, *Optical Pattern Recognition*. F. T. S. Yu and S. Jutamulia ed., 89–125, Cambridge University Press, Cambridge.
5. Ozaktas, H. M., Barshan, B., Mendolvic, D. and Oneral, L. 1994. Convolution, filtering, and multiplexing in fractional Fourier domains and their relation to chirp and wavelet transforms. *J. Opt. Soc. Amer.* A11: 547.
6. Ozaktas, H. M., Barshan, B. and Mendolvic, D. 1998. Convolution in fractional Fourier domains. *Opt. Rev.* 1: 15.
7. Lohmann, A. W., Zalevsky, Z. and Mendolvic, D. 1996. Synthesis pattern recognition filters for fractional Fourier processing. *Opt. Commun.* 128: 199.
8. Lohmann, A. W. and Mendlovic, D. 1997. Fractional joint transform correlator. *Appl. Opt.* 36: 7042.

A Numerical Calculation of Discrete Fresnel Diffraction

Consider an object $f(x,y)$ illuminated by a coherent light of wavelength λ, as shown in Fig. A.1. The observation plane is located at a distance d from the object plane, where an image sensor array is installed. The coordinates of the observation plane are chosen as (ξ, η).

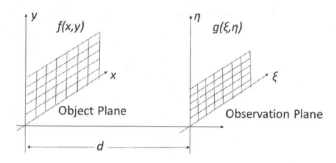

Figure A.1 Geometry for diffraction calculation.

Let us start with the Fresnel diffraction equation, Eq. (2.41) or Eq. (6.1).

$$g(\xi, \eta) = \frac{1}{i\lambda d}$$
$$\times \iint f(x,y) \exp\left\{i\frac{\pi}{\lambda d}\left[(\xi - x)^2 + (\eta - y)^2\right]\right\} dxdy. \quad (A.1)$$

This is rewritten as a convolution integral,

$$g(\xi, \eta) = f * h(\xi, \eta), \quad (A.2)$$

where

$$h(\xi, \eta) = \frac{1}{i\lambda d} \exp\left[i\frac{\pi}{\lambda d}(\xi^2 + \eta^2)\right]. \quad (A.3)$$

Fourier transforming Eq. (A.2) is given by

$$G(v_x, v_y) = F(v_x, v_y) \cdot H(v_x, v_y), \quad (A.4)$$

where

$$v_x = \frac{\xi}{\lambda d}, \qquad v_y = \frac{\eta}{\lambda d}$$

DOI: 10.1201/9781003121916-A

and

$$F(v_x, v_y) = \mathscr{F}[f(x,y)] \tag{A.5}$$

$$G(v_x, v_y) = \mathscr{F}[g(x,y)] \tag{A.6}$$

$$H(v_x, v_y) = \mathscr{F}[h(x,y)] = \exp[-i\pi\lambda d(v_x^2 + v_y^2)]. \tag{A.7}$$

The Fresnel diffraction equation, Eq.(A.1), is also rewritten as

$$g(\xi, \eta) = \frac{1}{i\lambda d} \exp\left[i\frac{\pi}{\lambda d}(\xi^2 + \eta^2)\right]$$
$$\times \iint f(x,y) \exp\left[i\frac{\pi}{\lambda d}(x^2 + y^2)\right] \exp\left[i\frac{2\pi}{\lambda d}(x\xi + y\eta)\right] dxdy. \tag{A.8}$$

There are three types of representation for Fresnel diffraction, as follows

1. Convolution approach
   According to Eqs. (A.2) and (A.4), the Fresnel diffraction is given by

$$g(\xi, \eta) = \mathscr{F}^{-1}\{\mathscr{F}[f] \cdot \mathscr{F}[h]\}. \tag{A.9}$$

It takes three Fourier transforms, or two Fourier transforms based on

$$g(\xi, \eta) = \mathscr{F}^{-1}\{\mathscr{F}[f] \cdot \exp[-i\pi\lambda d(v_x^2 + v_y^2)]\}. \tag{A.10}$$

2. Fresnel transform approach
   According to Eq. (A.8), the Fresnel diffraction is given by

$$g(\xi, \eta) = \frac{1}{i\lambda d} \exp\left[i\frac{\pi}{\lambda d}(\xi^2 + \eta^2)\right] \mathscr{F}\left\{f(x,y) \exp\left[i\frac{\pi}{\lambda d}(x^2 + y^2)\right]\right\}. \tag{A.11}$$

It takes only one Fourier transform.

3. Angular spectrum approach
   According to Eqs. (6.59), (6.61) and (6.63), the Fresnel diffraction based on the angular spectrum method is given by

$$u(x,y,d) = \iint_{-\infty}^{\infty} U(v_x, v_y, 0) \exp\left(i2\pi d\sqrt{\frac{1}{\lambda^2} - v_x^2 - v_y^2}\right)$$
$$\times \exp[i2\pi(v_x x + v_y y)] dv_x dv_y. \tag{A.12}$$

So we have

$$u(x,y,d) = \mathscr{F}\left\{\mathscr{F}[u(x,y,0)] \exp\left(i2\pi d\sqrt{\frac{1}{\lambda^2} - v_x^2 - v_y^2}\right)\right\}. \tag{A.13}$$

It takes two Fourier transforms.

Next, we discuss the sampling issues in the diffraction calculation. To calculate discrete Fourier transforms, all the data should satisfy the sampling theorem. At first, let the input image $f(x,y)$ satisfy the sampling theorem. The highest spatial frequency of the point spread function $h(\xi,\eta)$ limits the distance d between the object and observation planes.

The spatial frequencies of the point spread function $h(\xi,\eta)$ are[1]

$$v_\xi = \frac{1}{2\pi} \frac{\partial h(\xi,\eta)}{\partial \xi} = \frac{\xi}{\lambda d} \tag{A.14}$$

$$v_\eta = \frac{1}{2\pi} \frac{\partial h(\xi,\eta)}{\partial \eta} = \frac{\xi}{\lambda d}. \tag{A.15}$$

The sampling theory requests

$$\text{Max}\left[\left|\frac{\xi}{\lambda d}\right|\right] \leq \frac{1}{2\Delta} \tag{A.16}$$

$$\text{Max}\left[\left|\frac{\eta}{\lambda d}\right|\right] \leq \frac{1}{2\Delta}, \tag{A.17}$$

where Δ denotes the sampling period of the point spread function. Finally, we have the limit of the distance d,

$$d \geq \frac{N\Delta^2}{\lambda} \tag{A.18}$$

because the maximum spatial frequency is at the end corner of the point spread function, that is, ξ_{max} and $\eta_{max} = N\Delta/2$, where N denotes the sampling number of the image. For example, when $N = 1024$, $\lambda = 0.633\ \mu m$, $\Delta = 20$ vm, $d \geq 650$ mm. Under this condition, the convolution method of the discrete Fresnel diffraction calculation is performed without the aliasing error.

Otherwise, in the angular spectrum method of the discrete Fresnel diffraction calculation, the maximum frequency of the frequency response function $\exp\left(i2\pi d\sqrt{\frac{1}{\lambda^2} - v_x^2 - v_y^2}\right)$ depends strongly on the distance d. This means the limit of the distance d is very small as compared with that of the convolution method. The angular spectrum method in the digital version is valid in the case when the observation plane locates close to the object plane.

---

[1] It should be noted that the spatial frequency of a function $h(\xi) = \exp[i\phi(\xi)]$ is given by $v_\xi = 1/2\pi \cdot d\phi(\xi)/d\xi$.

B Numerical Calculation of Fresnel Hologram

Consider a diffuse object illuminated by a coherent light of the wavelength λ. The complex amplitude on the object surface is given by

$$f(x,y) = A(x,y)\exp[i\phi(x,y)], \tag{B.1}$$

where $A(x,y)$ and $\phi(x,y)$ denoted the amplitude and phase of the object.

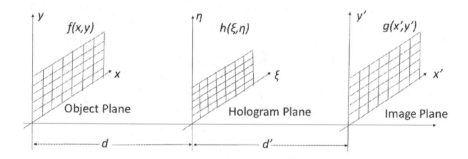

Figure B.1 Geometry for digital holography.

The geometry describing the numerical procedures of the digital holography is shown in Fig. B.1.

The hologram plane is located at a distance d from the object plane, where an image sensor array is installed. The coordinates of the hologram plane are chosen as (ξ, η). At a distance d' from the hologram plane whose coordinates are (x', y'), the image plane or the observation plane is set, where the intensity of the reconstructed image is obtained. Numerical processes, as mentioned later, give the complex amplitude $g(x', y')$ of the reconstruction image. Due to the Fresnel diffraction theory, according to Eq. (6.1), the complex amplitude in the hologram plane is given by

$$h(\xi, \eta) = \frac{\exp(ikd)}{i\lambda d}$$
$$\times \iint f(x,y)\exp\left\{i\frac{\pi}{\lambda d}\left[(\xi - x)^2 + (\eta - y)^2\right]\right\}dxdy \tag{B.2}$$

assuming the constant A in Eq. (2.41) to be unity. This is the convolution integral of the object $f(x,y)$ and the impulse response function $\exp\left\{i\frac{\pi}{\lambda d}\left[(\xi - x)^2 + (\eta - y)^2\right]\right\}$,

DOI: 10.1201/9781003121916-B

is rewritten as

$$h(\xi, \eta) = \frac{\exp(ikd)}{i\lambda d} \exp\left[i\frac{\pi}{\lambda d}(\xi^2 + \eta^2)\right]$$
$$\times \iint f(x,y) \exp\left[i\frac{\pi}{\lambda d}(x^2 + y^2)\right] \exp\left[-i\frac{2\pi}{\lambda d}(x\xi + y\eta)\right] dxdy. \quad (B.3)$$

Equation (B.3) is rewritten as

$$h(v_x, v_y) = \frac{\exp(ikd)}{i\lambda d} \exp[i\pi d\lambda(v_x^2 + v_y^2)]$$
$$\times \iint f(x,y) \exp\left[i\frac{\pi}{\lambda d}(x^2 + y^2)\right] \exp[-i2\pi(xv_x + y\mu_y)] dxdy, \quad (B.4)$$

where the spatial frequencies are defined as

$$v_x = \frac{\xi}{d\lambda}, \qquad v_y = \frac{\eta}{d\lambda}. \quad (B.5)$$

In practice, the object plane is sampled with the sampling period Δx and Δy. The hologram plane is also sampled by an image sensor array in the case of the digital holography, whose pixel spacing is $\Delta\xi$ and $\Delta\eta$ to the ξ and η directions, respectively. The discrete coordinates are given by

$$\xi = n\Delta\xi, \qquad n = 0, 1, ..., N \quad (B.6)$$
$$\eta = m\Delta\eta, \qquad m = 0, 1, ..., M. \quad (B.7)$$

According to Eq. (B.4), the discrete version of the spatial frequencies in the hologram plane are

$$v_x = n\Delta v_x, \qquad n = 0, 1, ..., N \quad (B.8)$$
$$v_y = m\Delta v_y, \qquad m = 0, 1, ..., M. \quad (B.9)$$

The sampling theory demands

$$\Delta x = \frac{1}{N\Delta v_x} = \frac{\lambda d}{N\Delta\xi} \quad (B.10)$$

$$\Delta y = \frac{1}{M\Delta v_y} = \frac{\lambda d}{M\Delta\eta}. \quad (B.11)$$

Next, assume that the reference wave in the hologram plane to be $r(\xi, \eta)$. The intensity distribution $h(\xi, \eta)$ is recorded as the digital hologram

$$h(\xi, \eta) = |F(\xi, \eta) + r(\xi, \eta)|^2. \quad (B.12)$$

To reconstruct a real image in the image plane from the hologram, the complex conjugate reference wave $r^*(\xi, \eta)$ is employed. To obtain the amplitude of the real reconstructed wavefront in the image plane, the inverse Fresnel diffraction is calculated.

$$g(x',y') = \exp\left[i\frac{\pi}{\lambda d}(x'^2 + y'^2)\right]$$
$$\times \iint h(\xi,\eta)r^*(\xi,\eta)\exp\left[i\frac{\pi}{\lambda d}(\xi^2 + \eta^2)\right]\exp\left[-i\frac{2\pi}{\lambda d}(x'\xi + y'\eta)\right]d\xi\,d\eta,$$
(B.13)

where d' is a set to be d, so as to obtain the focused image. Equation (B.13) is rewritten as

$$g(\mu_x, \mu_y) = \exp[i\pi\lambda d(\mu_x'^2 + \mu_y^2)]$$
$$\times \iint h(\xi,\eta)r^*(\xi,\eta)\exp\left[i\frac{\pi}{\lambda d}(\xi^2 + \eta^2)\right]\exp[-i2\pi(\xi\mu_x + \eta\mu_y)]d\xi\,d\eta,$$
(B.14)

where the spatial frequencies are defined as

$$\mu_x = \frac{x'}{d\lambda}, \qquad \mu_y = \frac{y'}{d\lambda}. \tag{B.15}$$

The discrete version of Eq. (B.14) is given by

$$g(n\Delta\mu_x, m\Delta\mu_y) = \exp[i\pi\lambda d(n^2\Delta\mu_x^2 + m^2\Delta\mu_y^2)]$$
$$\times \sum_{k=0}^{N-1}\sum_{l=0}^{M-1} h(k\Delta\xi, l\Delta\eta)r^*(k\Delta\xi, l\Delta\eta)$$
$$\times \exp\left[i\frac{\pi}{\lambda d}(k^2\Delta\xi^2 + l^2\Delta\eta^2)\right]\exp\left[-i2\pi\left(\frac{kn}{N} + \frac{lm}{M}\right)\right], \quad \text{(B.16)}$$

where

$$\Delta\mu_x = \frac{1}{N\Delta\xi} = \frac{\Delta x'}{\lambda d} \tag{B.17}$$

$$\Delta\mu_y = \frac{1}{M\Delta\eta} = \frac{\Delta y'}{\lambda d}. \tag{B.18}$$

From Eqs. (B.10) and (B.11) and Eqs. (B.17) and (B.18), we have

$$\Delta x = \Delta x', \qquad \Delta y = \Delta y'. \tag{B.19}$$

The sampling periods in the object plane and the image plane are equal to each other. According to Eqs. (B.17) and (B.18), the pixel size in the image plane is

$$\Delta x' = \frac{\lambda d}{N\Delta\xi}, \qquad \Delta y' = \frac{\lambda d}{M\Delta\eta}. \tag{B.20}$$

Solutions to Selected Problems

Chapter 1

1.2 velocity: $v = 4$ m/s, propagation direction: z, period: $T = 0.03125$ s, wavelength: $\lambda = 0.125$ m, wavenumber: $k = 50.24$ m^{-1}.

1.3 $v = c/\lambda = 2.998 \times 10^8 / 0.6328 \times 10^{-6} = 4.7 \times 10^{14}$.

Chapter 2

2.1 The beat effect occurs.

2.2

$$
\begin{aligned}
u(v_x) &= A' \int_{l/2-D/2}^{l/2+D/2} \exp(-i2\pi x v_x) dx + A' \int_{-l/2-D/2}^{-l/2+D/2} \exp(-i2\pi x v_x) dx \\
&= 2A' D \mathrm{sinc}(Dv_x) \cdot \cos(\pi l v_x).
\end{aligned}
$$

2.3

$$
\begin{aligned}
u(\omega, \phi) &= A' \int_{D_2/2}^{D_1/2} \int_0^{2\pi} \exp\left[-i\frac{k}{R}\rho\omega\cos(\theta - \phi)\right] \rho d\rho d\theta \\
&= A' \int_0^{D_1/2} \int_0^{2\pi} \exp\left[-i\frac{k}{R}\rho\omega\cos(\theta)\right] \rho d\rho d\theta \\
&\quad - A' \int_0^{D_2/2} \int_0^{2\pi} \exp\left[-i\frac{k}{R}\rho\omega\cos(\theta)\right] \rho d\rho d\theta \\
&= \pi A' \left[\left(\frac{D_1}{2}\right)^2 \frac{2J_1\left(\frac{kD_1}{2R}\omega\right)}{\frac{kD_1}{2R}\omega} - \left(\frac{D_2}{2}\right)^2 \frac{2J_1\left(\frac{kD_2}{2R}\omega\right)}{\frac{kD_2}{2R}\cdot\omega} \right].
\end{aligned}
$$

2.4 The diffracted wave from the (m,n) aperture is

$$
\begin{aligned}
u_{m,n}(v_x, v_y) &= A' \int_{-\infty}^{\infty} f(x - ma, y - nb) \exp[-i2\pi(xv_x + yv_y)] dxdy \\
&= A' \exp[-i2\pi(mav_x + nbv_y)] \cdot F(v_x, v_y),
\end{aligned}
$$

where

$$
F(v_x, v_y) = \int_{-\infty}^{\infty} f(x,y) \exp[-i2\pi(xv_x + yv_y)] dxdy,
$$

which gives the Fraunhofer diffraction from an aperture $f(x,y)$. The diffracted wave from $(2M+1) \times (2N+1)$ apertures is given by

$$u(v_x, v_y) = \sum_{m=-M}^{M} \sum_{n=-N}^{N} u_{m,n}(v_x, v_y)$$

$$= A' \frac{1 - \exp[-i2\pi(2M+1)av_x]}{1 - \exp(-i2\pi av_x)}$$

$$\times \frac{1 - \exp[-i2\pi(2N+1)bv_y]}{1 - \exp(-i2\pi bv_y)}$$

$$\times \exp[i2\pi(Mav_x + Nbv_y)] \times F(v_x, v_y).$$

Its intensity is

$$I(v_x, v_y) = A'^2 \left\{ \left[\frac{\sin[\pi(2M+1)av_x]}{\sin(\pi av_x)} \right]^2 \cdot \left\{ \left[\frac{\sin[\pi(2N+1)bv_y]}{\sin(\pi bv_y)} \right]^2 \right\} \right.$$

$$\times |F(v_x, v_y)|^2.$$

Chapter 3

3.1

$$f(x) = \frac{2}{\pi} + \sum_{n=1}^{\infty} (-1)^{n+1} \frac{4}{(4n^2 - 1)\pi} \cos(2\pi nx)$$

3.2

$$f(x) = \frac{1}{12} + \sum_{n=1}^{\infty} (-1)^n \frac{1}{n^2 \pi^2} \cos(2\pi nx)$$

3.3 (a)

$$\frac{i}{2\sqrt{\alpha}} \exp\left(-i\frac{\pi v^2}{2\alpha}\right)$$

(b)

$$\frac{\alpha}{\pi(a^2 + v^2)}$$

(c)

$$\frac{\alpha}{\alpha^2 + 4\pi^2(v - v_0)^2} + \frac{\alpha}{\alpha^2 + 4\pi^2(v + v_0)^2}.$$

3.4 (1) Convolution and Correlation

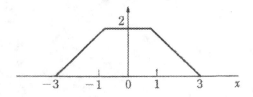

(2) Convolution and Correlation

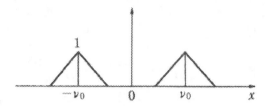

(3) Convolution

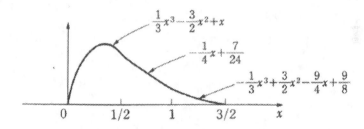

Correlation

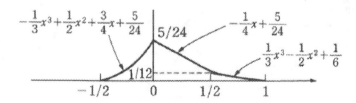

(4) Convolution and Correlation

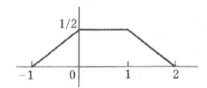

3.5 Direct method:

$$\exp(-\alpha x^2) * \exp(-\beta x^2)$$

$$= \int_{-\infty}^{\infty} \exp[-\alpha x'^2 - \beta (x'-x)^2] dx'$$

$$= \int_{-\infty}^{\infty} \exp\left[-(\alpha+\beta)\left(x' - \frac{\beta x'}{\alpha+\beta}\right)^2 + \frac{(\beta x)^2}{\alpha+\beta} - \beta x^2\right] dx'$$

$$= \exp\left(-\frac{\alpha\beta}{\alpha+\beta}x^2\right) 2 \int_0^{\infty} \exp[-(\alpha+\beta)x'^2] dx'$$

$$= \sqrt{\frac{\pi}{\alpha+\beta}} \exp\left(-\frac{\alpha\beta}{\alpha+\beta}x^2\right).$$

Fourier transform method:

Since $\exp(-\alpha x^2) = \exp\left[-\pi\left(\sqrt{\frac{\alpha}{\pi}}x\right)^2\right]$,

$$\exp(-\alpha x^2) \Longleftrightarrow \sqrt{\frac{\pi}{\alpha}} \exp\left[-\pi\left(\sqrt{\frac{\pi}{\alpha}}v\right)^2\right]$$

$$\exp(-\beta x^2) \Longleftrightarrow \sqrt{\frac{\pi}{\beta}} \exp\left[-\pi\left(\sqrt{\frac{\pi}{\beta}}v\right)^2\right]$$

$$\mathscr{F}[\exp(-\alpha x^2) * \exp(-\beta x^2)] = \frac{\pi}{\sqrt{\alpha\beta}} \exp\left[-\pi\left(\sqrt{\frac{\pi(\alpha+\beta)}{\alpha\beta}}v\right)^2\right].$$

Its inverse Fourier transform is

$$\exp(-\alpha x^2) * \exp(-\beta x^2)$$

$$= \frac{\pi}{\sqrt{\alpha\beta}} \sqrt{\frac{\alpha\beta}{\pi(\alpha+\beta)}} \exp\left[-\pi\left(\sqrt{\frac{\alpha\beta}{\pi(\alpha+\beta)}}x\right)^2\right]$$

$$= \sqrt{\frac{\pi}{\alpha+\beta}} \exp\left(-\frac{\alpha\beta}{\alpha+\beta}x^2\right).$$

3.6 (1) Since

$$\int_{-\infty}^{\infty} f(x)\delta(x)dx = f(0)$$

and

$$\int_{-\infty}^{\infty} f(ax)\delta(x)dx = \int_{-\infty}^{\infty} f\left(\frac{x}{a}\right)\delta(x)\frac{dx}{|a|} = f(0),$$

we have

$$\delta(ax) = \frac{1}{|a|}\delta(x).$$

(2) From Eq. (3.82) with $T = 1$, we have

$$\text{comb}(x) = 1 + 2\sum_{n=1}^{\infty} \cos(2\pi nx)$$

$$\mathscr{F}[\text{comb}(x)] = \delta(v_x) + 2\sum_{n=1}^{\infty} \mathscr{F}[\cos(2\pi nx)]$$

$$= \delta(v_x) + \sum_{n=1}^{\infty} [\delta(v_x - n) + \delta(v_x + n)]$$

$$= \sum_{n=-\infty}^{\infty} \delta(v_x - n)$$

$$= \text{comb}(v_x).$$

(3) Combination of (1) and (2).

$$\mathscr{F}[\text{comb}(ax)] = \sum_{n=-\infty}^{\infty} \delta(ax - n) = \sum_{n=-\infty}^{\infty} \delta[a(x - n/a)]$$

$$= \frac{1}{|a|}\sum_{n=-\infty}^{\infty} \delta(x - n/a)$$

3.7 (1) Since

$$\mathscr{F}[\exp(-|x|)] = \frac{2}{1 + (2\pi v)^2},$$

we have

$$\int_{-\infty}^{\infty} \exp(-2|x|)dx = \int_{-\infty}^{\infty} \left[\frac{2}{1 + (2\pi v)^2}\right]^2 dv$$

and because

$$\int_{-\infty}^{\infty} \exp(-2|x|)dx = 1,$$

$$\int_{-\infty}^{\infty} \frac{dv_x}{(1 + v_x^2)^2} = \frac{\pi}{2}.$$

(2) Since

$$\mathscr{F}[\text{rect}(x)] = \text{sinc}(v),$$

we have

$$\int_{-\infty}^{\infty} [\text{rect}(x)]^2 dx = \int_{-\infty}^{\infty} [\text{sinc}(v)]^2 dv$$

$$\int_{-\infty}^{\infty} [\text{rect}(x)]^2 dx = \int_{-1/2}^{1/2} dx = 1$$

$$\int_{-\infty}^{\infty} \text{sinc}^2(x)dx = 2\int_{0}^{\infty} \frac{\sin^2 \pi v}{(\pi v)^2} dv.$$

Finally we have

$$\int_0^\infty \frac{\sin^2 \pi v}{(\pi v)^2} dx = \frac{1}{2}.$$

Chapter 4

4.1 Input–output response in audio equipment, response function in RC circuits, imaging system based on paraxial optics, vibration response of buildings for minute amplitude modulation, and so on.

4.2

$$\mathscr{F}[f(x)] = \left\{ \delta(v_x) + \frac{1}{2}[\delta(v_x - v_1) + \delta(v_x + v_1)] \right\}$$

$$* \left\{ \delta(v_x) + \frac{1}{2}[\delta(v_x - v_2) + \delta(v_x + v_2)] \right\}$$

$$= \delta(v_x) + \frac{1}{2}[\delta(v_x - v_1) + \delta(v_x + v_1) + \delta(v_x - v_2) + \delta(v_x + v_2)$$

$$+ \frac{1}{4}[\delta(v_x - v_1 - v_2) + \delta(v_x - v_1 + v_2) + \delta(v_x + v_1 - v_2)$$

$$+ \delta(v_x + v_1 + v_2)].$$

This means that the system frequency response should be 1 at the frequency $v_x = 0, \pm v_1, \pm v_2, v_1 + v_2, v_1 - v_2, -v_1 + v_2, -(v_1 + v_2)$.

Chapter 5

5.6 Direct method: N^2, FFT method: $(2N \log N) \times 2 + N$.

5.7 FFT is an algorithm to calculate the discrete Fourier transform with complex values. This means that the real Fourier transform can be performed with the zero-valued imaginary part. An ingenious algorithm is proposed, in which the imaginary part is employed efficiently.

Consider the discrete Fourier transform of a real sequence $f(n), (n = 0, 1, \cdots, N-1)$,

$$F(l) = \sum_{n=0}^{N-1} f(n) \exp\left(-\frac{i2\pi nl}{N}\right)$$

$$= \sum_{m=0}^{N/2-1} \left\{ f(2m) \exp\left(-\frac{i4\pi ml}{N}\right) \right.$$

$$\left. + f(2m+1) \exp\left[-\frac{i2\pi(2m+1)l}{N}\right] \right\}$$

$$= F_1(2m) + \exp\left(-\frac{i2\pi l}{N}\right) F_2(2m+1), \qquad \text{(B.21)}$$

where $l = 0, 1, \cdots, N-1$, and $F_1(2m)$ and $F_2(2m+1)$ are the discrete Fourier transform of the even and odd number sample $f(2m)$ and $f(2m+1)$, respectively. This means that the discrete Fourier transform of a real sequence $f(n)$ is calculated by two discrete Fourier transforms $F_1(2m)$ and $F_2(2m+1)$ and Eq. (B.21).

Another method to calculate two real functions is as follows. Consider 2 real sequences $g_1(n)$ and $g_2(n)$, $(n = 0, 1, \cdots, N-1)$. A complex sequence

$$h(n) = g_1(n) + ig_2(n)$$

is defined. Its Fourier transform is

$$H(l) = H_r(l) + iH_j(l),$$

where $H_r(l)$ and $H_i(l)$ are the real and imaginary part of $H(l)$.

$$G_1(l) = \frac{H_r(l) + H_r(N-l)}{2} + i\left[\frac{H_i(l) - H_i(N-l)}{2}\right]$$

$$G_2(l) = \frac{H_i(l) + H_i(N-l)}{2} + i\left[\frac{H_r(l) - H_r(N-l)}{2}\right]$$

Chapter 6

6.1

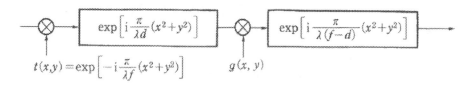

$$t(x,y) = \exp\left[-i\frac{\pi}{\lambda f}(x^2+y^2)\right] \qquad g(x,y)$$

$$h(x,y) = \left[\left\{\exp\left[-i\frac{\pi}{\lambda f}(x^2+y^2)\right] * \exp\left[i\frac{\pi}{\lambda d}(x^2+y^2)\right]\right\}g(x,y)\right]$$

$$* \exp\left[i\frac{\pi}{\lambda(f-d)}(x^2+y^2)\right]$$

$$= \left[\iint \exp\left[-i\frac{\pi}{\lambda f}(x_1^2+y_1^2)\right]\right.$$

$$\exp\left\{i\frac{\pi}{\lambda d}[(x-x_1)^2+(y-y_1)^2]\right\}dx_1dy_1$$

$$\left.\times g(x,y)\right] * \exp\left[i\frac{\pi}{\lambda(f-d)}(x^2+y^2)\right]$$

$$= i\frac{\lambda fd}{f-d}\exp\left[i\frac{\pi}{\lambda(f-d)}(x^2+y^2)\right]$$

$$\times G\left[\frac{x}{\lambda(f-d)}, \frac{y}{\lambda(f-d)}\right]$$

6.2

$$h(x,y) = \iiiint g(x_1,y_1)\exp\left[i\frac{\pi}{2\lambda f}(x_2-x_1)^2+(y_2-y_1)^2\right]$$

$$\times \exp\left[-i\frac{\pi}{\lambda f}(x_2^2+y_2^2)\right]$$

$$\times \exp\left[i\frac{\pi}{\lambda f}(x-x_2)^2+(y-y_2)^2\right]dx_1dy_1dx_2dy_2$$

$$= \iiiint g(x_1,y_1)\exp\left\{i\frac{\pi}{2\lambda f}\left[\frac{2f-d}{d}(x_2^2+y_2^2)\right.\right.$$

$$-2\left(x_1+\frac{2f}{d}x\right)x_2 - 2\left(y_1+\frac{2f}{d}y\right)y_2 + x_1^2+y_1^2$$

$$\left.\left.+\frac{2f}{d}(x^2+y^2)\right]\right\}dx_1dy_1dx_2dy_2.$$

If $2f = d$,

$$
\begin{aligned}
h(x,y) &= \iiiint g(x_1,y_1)\exp\Big\{i\frac{\pi}{2\lambda f}\Big[\\
&\quad -2(x_1+x)x_2 - 2(y_1+y)y_2 + x_1^2 + y_1^2 \\
&\quad + x^2 + y^2\Big]\Big\}dx_1 dy_1 dx_2 dy_2 \\
&= \iint g(x_1,y_1)\delta(x_1+x,y_1+y) \\
&\qquad \exp\Big[i\frac{\pi}{2\lambda f}(x_1^2+y_1^2+x^2+y^2)\Big]dx_1 dy_1 \\
&= g(-x,-y)\cdot\exp\Big[i\frac{\pi}{2\lambda f}(x^2+y^2)\Big].
\end{aligned}
$$

6.3 (a) Combination of the transmittances of lenses is

$$
\begin{aligned}
&\exp\Big[-i\frac{\pi}{\lambda f_1}(x^2+y^2)\Big]\cdot\exp\Big[-i\frac{\pi}{\lambda f_2}(x^2+y^2)\Big]. \\
&= \exp\Big[-i\frac{\pi}{\lambda}\cdot\frac{f_1+f_2}{f_1 f_2}(x^2+y^2)\Big]
\end{aligned}
$$

The focal length of the lens pair is

$$
\frac{1}{f} = \frac{f_1+f_2}{f_1 f_2} = \frac{1}{f_1} + \frac{1}{f_2}.
$$

(b)

$$
\begin{aligned}
&\Big\{\exp\Big[-i\frac{\pi}{\lambda f_1}(x^2+y^2)\Big] * \exp\Big[i\frac{\pi}{\lambda d}(x^2+y^2)\Big]\Big\} \\
&\quad \times \exp\Big[-i\frac{\pi}{\lambda f_2}(x^2+y^2)\Big] \\
&= \iint \exp\Big[-i\frac{\pi}{\lambda f_1}(x_1^2+y_1^2)\Big] \\
&\quad \times \exp\Big\{i\frac{\pi}{\lambda d}[(x-x_1)^2+(y-y_1)^2]\Big\}dx_1 dy_1 \\
&\quad \times \exp\Big[-i\frac{\pi}{\lambda f_2}(x^2+y^2)\Big] \\
&= \iint \exp\Big\{-i\frac{\pi}{\lambda f_1}\Big(\frac{d-f_1}{d}\Big)\Big[\Big(x_1+\frac{f_1 x}{d-f_1}\Big)^2 \\
&\quad + \Big(y_1+\frac{f_1 y}{d-f_1}\Big)^2\Big] + i\frac{\pi(x^2+y^2)}{(d-f_1)\lambda}\,dx_1 dy_1 \\
&\quad \times \exp\Big[-i\frac{\pi}{\lambda f_2}(x^2+y^2)\Big] \\
&\propto \exp\Big[-i\frac{\pi}{\lambda}(x^2+y^2)\Big(\frac{1}{f_2}-\frac{1}{d-f_1}\Big)\Big].
\end{aligned}
$$

6.5

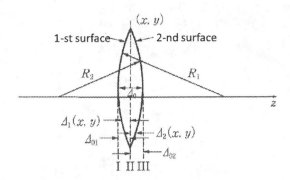

Consider a lens with the radius of curvatures of R_1 and R_2, as shown in the figure. The thickness of the lens is $\Delta(x,y)$ and its central part Δ_0, which is the distance between the I and III plane. The phase change of the lens is given by

$$\phi(x,y) = \frac{2\pi}{\lambda} n\Delta(x,y) + \frac{2\pi}{\lambda}[\Delta_0 - \Delta(x,y)]$$

$$= \frac{2\pi}{\lambda}\Delta_0 + \frac{2\pi}{\lambda}[(n-1)\Delta(x,y)].$$

The lens is divided into two parts, the front part and back part, and its distances from the center are denoted by Δ_{01} and Δ_{02}. We have $\Delta_0 = \Delta_{01} + \Delta_{02}$. As shown in the figure, the distances from the first and second surfaces to the center of the lens are denoted by $\Delta_1(x,y)$ and $\Delta_2(x,y)$, and we have $\Delta(x,y) = \Delta_1(x,y) + \Delta_2(x,y)$.

Due to the geometrical relation,

$$\Delta_1(x,y) = \Delta_{01} - \left(R_1 - \sqrt{R_1^2 - x^2 - y^2}\right)$$

$$\approx \Delta_{01} - \frac{x^2 + y^2}{2R_1}$$

$$\Delta_2(x,y) \approx \Delta_{02} - \frac{x^2 + y^2}{2R_2}.$$

We have

$$\Delta(x,y) = \Delta_0 - \frac{x^2 + y^2}{2}\left(\frac{1}{R_1} + \frac{1}{R_2}\right).$$

Finally, the transmittance of the lens is

$$t(x,y) = \exp\left(i\frac{2\pi}{\lambda}n\Delta_0\right)\exp\left[-i\frac{2\pi}{\lambda}(n-1)\left(\frac{1}{R_1} + \frac{1}{R_2}\right)\frac{x^2 + y^2}{2}\right]$$

$$= \exp\left(i\frac{2\pi}{\lambda}n\Delta_0\right)\exp\left[-i\frac{\pi}{\lambda f}(x^2 + y^2)\right],$$

where

$$\frac{1}{f} = (n-1)\left(\frac{1}{R_1} + \frac{1}{R_2}\right).$$

Since the phase change due to the thickness of the lens Δ_0 is usually neglected, Eq. (6.13) is obtained.

6.6

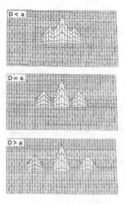

6.8 The spectrum of the object is shown in Fig. (a). In the case of the coherent imaging (b) and if the cut-off frequency $v_c > v_0$, the ideal image is obtained without any attenuation.

In the incoherent imaging (c), the ideal imaging is performed when $v_c \geq 2v_0$. The blurred image is formed, due to the attenuation in the spatial frequency, when $2v_0 > v_c \geq v_0$. No image is obtained when $v_0 > v_c$.

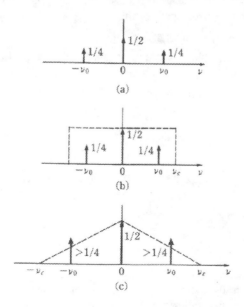

Chapter 7

7.1 The holography is an image recoding technique in which the complex amplitude from the object is recorded by interfering the object and reference waves. The holography is a kind of perfect photographic method, because the amplitude and phase of the object can be recorded. On the other hand, conventional photography can record only the intensity of the image without the phase information of the object. Because of the phase recording ability of the holography, recording and reconstruction of 3-D objects are able to be performed. Interferometry for diffused objects is also introduced, which is applied to deformation measurement and vibration analysis of many mechanical parts.

Many other applications of the holography have been described. See for examples, Hariharan, P. 1996. *Optical Holography*, Cambridge University Press. Kreis, T. 2006. *Handbook of Holographic Interferometry, Optical and Digital Methods*, John Wiley & Sons.

7.2 The computer-generated holography enables the reconstruction and display of artificial objects, generation of optical wavefront, which is difficult to synthesize, and so on. The practical applications are;

(1) Display of 3-D artificial object

Display in computer graphics and simulation, 3-D display of results in CAD, some models in automobile models, and so on.

(2) Interference prototype

In interferometric testing of aspherical lenses and mirrors, reference wavefronts with ideal shapes are generated by CGHs. The resultant interferometric fringe pattern shows the error of the test object.

(3) Holographic optical element (HOE)

Holographic optical elements using CGHs are used in laser beam shaping, optical beam scanning and optical devices in optical interconnects.

(4) Complex spatial filtering

Described in Sec. 8.1.

Chapter 8

8.1 Consider the Fourier spectra $F(v_x, v_y)$ and $G(v_x, v_y)$ of the object $f(x,y)$ and $g(x,y)$, respectively. The code transform filter from $f(x,y)$ to $g(x,y)$ is

$$\frac{G(v_x, v_y)}{F(v_x, v_y)}.$$

8.2 The transform from the cartesian coordinates to polar coordinates is the rotation invariant transform. The scale and rotation invariant transforms have been discussed by many authors. For example, (1): Casasent, D. and Psaltis, D.

1976. Position, rotation and scale invariant optical correlation, *Appl Opt.*, 15: 1795 (2): Saito, Y., Komatu, S. and Ohzu, H. 1983. Scale and rotation invariant real time optical correlator using computer generated hologram, *Opt Commun.* 47: 8.

Chapter 9

9.1 Consider Fourier transforms $V(v)$ and $\hat{V}(v)$ of $v(t)$ and $\hat{v}(t)$, respectively. Because

$$\hat{V}(v) = -i\,\text{sgn}(v)V(v),$$

we have

$$-i\,\text{sgn}(v)\hat{V}(v) = -i\,\text{sgn}(v) \cdot [-i\,\text{sgn}V(v)] = -V(v).$$

Therefore,

$$\hat{\hat{v}}(t) = -v(t).$$

9.2 Fourier transform of $v(t) = m(t)\cos(2\pi v_0 t)$ is

$$V(v) = M(v) * \frac{\delta(v - v_0) + \delta(v + v_0)}{2}.$$

Its Hilbert transform is

$$\hat{V}(v) = -i\,\text{sgn}(v) \cdot \left[M(v) * \frac{\delta(v - v_0) + \delta(v + v_0)}{2}\right]$$

$$= -i\,\text{sgn}(v)\left[\frac{M(v - v_0)}{2} + \frac{M(v + v_0)}{2}\right]$$

$$= \frac{-iM(v - v_0)}{2} + \frac{iM(v + v_0)}{2}$$

$$= \frac{M(v - v_0) - M(v + v_0)}{2i} = M(v) * \frac{\delta(v - v_0) - \delta(v + v_0)}{2i}.$$

Finally, its inverse Fourier transform is

$$\hat{v}(t) = m(t)\sin(2\pi v_0 t).$$

9.3 Consider Fourier transforms $V(v)$ and $\hat{V}(v)$ of $v(t)$ and $\hat{v}(t)$, respectively. Because

$$\hat{V}(v) = -i\,\text{sgn}(v)V(v).$$

$V(v)$ and $\hat{V}(v)$ are different only in the phase, and therefore $|\hat{V}(v)|^2 = |V(v)|^2$. Their power spectra are the same.

9.4 Because Hilbert transform of the signal $v(t) = m(t)\cos(2\pi\mu_0 t)$ is $\hat{v}(t) = m(t)\sin(2\pi\mu_0 t)$, the analytic signal of $v(t)$ is

$$z(t) = v(t) + i\hat{v}(t).$$

Then,

$$z(t) = m(t)\cos(2\pi v_0 t) + im(t)\sin(2\pi v_0 t)$$
$$= m(t)\exp(i2\pi v_0 t).$$

The phase of the analytic signal is given by

$$\phi_z(t) = \frac{\hat{v}(t)}{v(t)}$$

called the instantaneous phase. The time differential of the instantaneous phase

$$2\pi v_z = \frac{\mathrm{d}\phi_z(t)}{\mathrm{d}t}$$

gives the instantaneous frequency v_z.

Chapter 10

10.1
$$V = \frac{5-2}{5+2} = \frac{3}{7}$$
$$V = \frac{2\sqrt{I_1 I_2}}{I_1 + I_2} = \frac{3}{5}$$

Finally, we have $\gamma_{12} = 0.46$.

10.2 The aliasing error. A sufficient number of samples which satisfy the sampling theory.

10.3 Let the thickness of dispersive material be d. The phase change is

$$\phi(\sigma) = 2\pi\sigma d[n(\sigma) - 1],$$

where σ denotes the wavenumber. Its interferogram is

$$I(h) = 2\int_0^\infty B(\sigma)\cos[2\pi\sigma h + \phi(\sigma)]\mathrm{d}\sigma.$$

This means that the interference fringe $I_\sigma(h)$ for the wavenumber σ is shifted depending on $\phi(\sigma)$, and therefore the symmetry of the interferometry to h is lost.

10.4 The interference fringe $I(x,y,l)$ includes the higher-order terms, like $\cos(2\pi n l/\lambda)$ and $\sin(2\pi n l/\lambda)$. This means Eq. (10.53) is not valid. The calculated phase includes the effects of the higher harmonic terms and therefore vibrating terms with higher frequencies appear.

10.5 Refer to Creath, K. 1988. *Progress in Optics*, Wolf, E. ed. Vol. XXVI, 351. North-Holland, Amsterdam

10.6 In the heterodyne interferometry, the phase of the beat between two adjacent frequency v_1 and v_2 is measured, and its intensity is

$$I(x,y,t) = a(x,y) + b(x,y)\cos[\phi(x,y) - 2\pi(v_1 - v_2)t].$$

On the other hand, the phase-shifting interferometry, the fringe intensity is

$$I(x,y,l) = a(x,y) + b(x,y)\cos\left[\phi(x,y) - \frac{2\pi l}{\lambda}\right].$$

In the heterodyne interferometry, the phase of the fringes is modulated temporally,

$$\theta = -2\pi(v_1 - v_2).$$

while in the phase-shifting interferometry,

$$\theta = -2\pi\frac{l}{\lambda}$$

Both interferometric techniques are the same because the interferometric fringes are modulated by the outer phase θ.

In the case, of the Fourier transform fringe analysis, we have from Eq. (10.59),

$$\theta = -\alpha x.$$

In this case, the phase is modulated spatially.

Index

3-D spectrum, 98

Airy disc, 29, 95, 119
aliasing error, 59
amplitude, 3
analytic signal, 149
angular spectrum method, 96
angular frequency, 4
angular spectrum, 96
aperture function, 23
apodization, 119

band-limited signal, 58, 72
Bartlett window function, 74

CGH, 107
    cell-oriented—, 107
    Lohmann type—, 107
    point-oriented, 109
circular function, 53
coherence, 16, 155
    spatial—, 159
    temporal—, 157
coherent imaging, 89
comb function, 53
complex amplitude, 7, 8
computer generated holography, 107
    binary—, 107
    cell-oriented—, 107
    Lohmann type—, 107
    point-oriented, 109
conjugate image, 106
convolution, 49
convolution theorem, 51
coordinate transform, 134
correlation, 49
correlation theorem, 51
CT, 140

degree of coherence, 16
delta function, 47
DFT, 73
differentiation filter, 117
diffraction, 18
    Fraunhofer—, 25
    Fresnel—, 23, 85

diffraction grating, 29
digital holography, 111
discrete Fourier transform, 73

eigenfunction, 67
eigenvalue, 67
etendue, 162

Fabry-Perot interferometer, 18
fast Fourier transform, 71
FDOCT, 179
FFT, 71, 75
    interpolation in —, 80
    numerical calculation by—, 78
    zero-padding in —, 80
filter
    Laplacian—, 117
    differentiation, 117
    high-pass—, 116
    low-pass—, 116
    matched—, 120
    phase-contrast—-, 118
Fizeau interferometer, 18
Fourier domain OCT, 179
Fourier integral, 26
Fourier series, 33
Fourier transform, 26, 33, 41
    discrete—, 73
    fast—, 71
Fourier transform fringe analysis, 167
Fourier transform spectroscopy, 160
fractional Fourier transform, 197
    —correlator, 202
    —joint fractional Fourier transform
        correlator, 204
    —matched filter, 202
    optical system of —, 199, 200
    WDF of —, 198
    —Wiener filter, 201
Fraunhofer diffraction, 25
    —circular aperture, 28
    —rectangular aperture, 27
frequency, 4, 34
frequency response, 93
frequency response function, 66
Fresnel diffraction, 23, 85, 211

fringe analysis
    Fourier transform—, 167
    Hilbert Transform—, 168

Gaussian function, 53
generalized Hamming window
        function, 74

Hamming window function, 75
hanning window function, 74
Heaviside step function, 47
high-pass filter, 116
Hilbert transform, 150
Hilbert Transform fringe analysis,
        168
hologram, 105
holography, 105
    computer generated—, 107
    conventional—l, 105
    digital—, 111
Huygens principle, 21

imaging
    coherent—, 89
    incoherent—, 93
impulse response, 65, 85
in-phase component, 152
incoherent imaging, 93
intensity, 8
interference, 13
interferometer, 18
    Fabry-Perot—, 18
    Fizeau—, 18
    Mach-Zehnder—, 18
    Michelson—, 18
    Twyman-Green—, 18
interpolation in DFT, 82

Jamin interferometer, 18
joint transform correlator, 130

kinoform, 111

Laplacian filter, 117
lens, 86
    Fourier transform by—, 86
    time—, 191
linear system, 63, 64
linearity, 44
low-pass filter, 116

Mach-Zehnder interferometer, 18
matched filter, 120
Mellin transform, 136
Michelson interferometer, 18
Michelson stellar interferometer, 160

negative frequency, 148
numerical calculation, 78

object wave, 105
OCT, 179
optical addition, 131
optical coherence tomography, 179
optical correlator, 128, 129
        space-integral—, 129
        time-integral—, 129
optical distance, 13
optical subtraction, 131
optimum filter, 123, 125
        —additive noise, 123
        —multiplicative noise, 125
optimum polynomial function, 39
orthogonal polynomial, 40

Parseval theorem, 51
phase, 3
phase difference, 14
phase shift interferometry, 163
phase-contrast filter, 118
plane wave, 3
point spread function, 85
principle of superposition, 8
projection-slice theorem, 142
propagation constant, 3
Pulse joint transform correlator, 177
Python, 78

quadrature component, 152

Radon transform, 142
Rayleigh criterion, 95
rectangular function, 52
rectangular window function, 74
reference wave, 105
resolving power, 95

sampling theory, 56
scalar wave, 9
shift invariant, 65
shift theorem, 46

shift-invariant system, 64
sign function, 53
similarity, 44
sinc function, 52
spatial frequency filtering, 115
spatial light modulator, 127
Spatio-temporal JTC
    schematic setup, 177
spatio-temporal WDF, 188
    — of dispersion, 191
    — of grating, 192
    — of lens, 190
    — of propagation, 191
    matrix representation of —, 193
    — of phase modulator, 191
    — of time lens, 191
spectrum analyzer, 126
spherical wave, 6
super resolution, 119
symmetricity, 44

time lens, 191
triangular function, 53
Twyman-Green interferometer, 18

uncertainty principle, 158

van Cittert-Zernike theorem, 160
vector wave, 9
visibility, 15

wave, 1
wave equation, 2
wave length, 3
wave number, 3
wave number vector, 5
wavefront, 4
wavelet transform, 138
WDF
    — of fractional Fourier transform, 198
    —of lens, 187
    —of propagation, 187
    spatio-temporal—, 188
WDF(Wigner distribution function), 185
Wiener filter, 124
Wiener-Khinchine theorem, 158
Wigner distribution function(WDF), 185
window function, 73
    Bartlett—, 74
    generalized Hamming—, 74
    Hamming—, 75
    hanning—, 74
    rectangular—, 74

X-ray computer tomography, 140
    2-D Fourier transform method in —, 142

Young's Experiment, 16

zero-padding, 80

Printed in the United States
by Baker & Taylor Publisher Services